Collins

MW00697507

NEW MATHS FRAMEWORKING

Matches the revised KS3 Framework

Kevin Evans, Keith Gordon, Trevor Senior, Brian Speed

William Collins' dream of knowledge for all began with the publication of his first book in 1819. A self-educated mill worker, he not only enriched millions of lives, but also founded a flourishing publishing house. Today, staying true to this spirit, Collins books are packed with inspiration, innovation and practical expertise. They place you at the centre of a world of possibility and give you exactly what you need to explore it.

Collins. Freedom to teach.

Published by Collins
An imprint of HarperCollins*Publishers*
77–85 Fulham Palace Road
Hammersmith
London
W6 8JB

Browse the complete Collins catalogue at
www.collinseducation.com

© HarperCollins*Publishers* Limited 2008

10 9 8

ISBN 978-0-00-726800-9

Keith Gordon, Kevin Evans, Brian Speed and Trevor Senior assert their moral rights to be identified as the authors of this work.

All rights reserved. No part of this publication may be reproduced, stored in a retrieval system, or transmitted in any form or by any means, electronic, mechanical, photocopying, recording or otherwise, without the prior written permission of the Publisher or a licence permitting restricted copying in the United Kingdom issued by the Copyright Licensing Agency Ltd., 90 Tottenham Court Road, London W1T 4LP.

Any educational institution that has purchased one copy of this publication may make unlimited duplicate copies for use exclusively within that institution. Permission does not extend to reproduction, storage within a retrieval system, or transmittal in any form or by any means, electronic, mechanical, photocopying, recording or otherwise, of duplicate copies for loaning, renting or selling to any other institution without the permission of the Publisher.

British Library Cataloguing in Publication Data
A Catalogue record for this publication is available from the British Library.
Commissioned by Melanie Hoffman and Katie Sergeant
Project management by Priya Govindan
Covers management by Laura Deacon
Edited by Karen Westall
Proofread by Amanda Dickson
Design and typesetting by Newgen Imaging
Design concept by Jordan Publishing Design
Covers by Oculus Design and Communications
Illustrations by Tony Wilkins and Newgen Imaging
Printed and bound by Printing Express, Hong Kong
Production by Simon Moore

Every effort has been made to trace copyright holders and to obtain their permission for the use of copyright material. The authors and publishers will gladly receive any information enabling them to rectify any error or omission in subsequent editions.

MIX
Paper from
responsible sources
FSC™ C007454

FSC™ is a non-profit international organisation established to promote the responsible management of the world's forests. Products carrying the FSC label are independently certified to assure consumers that they come from forests that are managed to meet the social, economic and ecological needs of present and future generations, and other controlled sources.

Find out more about HarperCollins and the environment at
www.harpercollins.co.uk/green

Introduction

Welcome to *New Maths Frameworking*!

New Maths Frameworking Year 8 Practice Book 3 has hundreds of levelled questions to help you practise Maths at Levels 6-7. The questions correspond to topics covered in Year 8 Pupil Book 3 giving you lots of extra practice.

These are the key features:

- **Colour-coded National Curriculum levels** for all the questions show you what level you are working at so you can easily track your progress and see how to get to the next level.

- **Functional Maths** is all about how people use Maths in everyday life. Look out for the Functional Maths icon **FM** which shows you when you are practising your Functional Maths skills.

Contents

CHAPTER 1 Number and Algebra **1** _____ 1

CHAPTER 2 Geometry and Measures **1** _____ 7

CHAPTER 3 Statistics **1** _____ 12

CHAPTER 4 Number **2** _____ 17

CHAPTER 5 Algebra **2** _____ 21

CHAPTER 6 Geometry and Measures **2** _____ 26

CHAPTER 7 Algebra **3** _____ 31

CHAPTER 8 Number **3** _____ 34

CHAPTER 9 Geometry and Measures **3** _____ 41

CHAPTER 10 Algebra **4** _____ 46

CHAPTER 11 Statistics **2** _____ 50

CHAPTER 12 Number **4** _____ 54

CHAPTER 13 Algebra **5** _____ 58

CHAPTER 14 Solving Problems _____ 63

CHAPTER 15 Geometry and Measures **4** _____ 67

CHAPTER 16 Statistics **3** _____ 75

Practice

1A Multiplying and dividing negative numbers

1 Calculate these.

   a $5 - 9$ **b** $2 - 6 + 3$ **c** $-2 + -5$ **d** $9 - -7$ **e** $-3 - +5 - 4$

2 **i**   Add each number to the number on its left. Write down your answers.
    ii Subtract each number from the number on its left. Write down
       your answers.

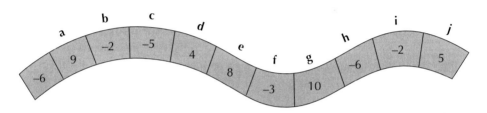

3 Calculate these.

a 5×-3	**b** -8×-5	**c** -3×10	**d** -2×-2
e 14×-3	**f** $-4 \times -2 \times 5$	**g** $2 \times -3 \times -1$	**h** $-10 \times 6 \times -10$
i $16 \div -2$	**j** $-6 \div -6$	**k** $-9 \div 3$	**l** $12 \div -3$
m $-100 \div 10$	**n** $20 \div -2 \div -5$	**o** $-36 \div -3 \div -4$	**p** $40 \div -5 \div -2$

4 Copy and complete these chains of calculations.

a

    -2 $\times 3$ ◯ $\div 2$ ◯ $+ -5$ ◯ $+3$ ◯

b

    12 $\div -2$ ◯ $+ -4$ ◯ $\times -3$ ◯ -50 ◯

c

    8 $+ -6$ ◯ $- -2$ ◯ $\times -3$ ◯ $+ -6$ ◯

5 **a** Write down five calculations involving $\times$ that give the answer -6.
    b Write down five calculations involving $\div$ that give the answer -5.

6 Calculate these.

a $(-8)^2$	**b** $4 + -2 \times 7$	**c** $-14 \div 2 + 5$
d $2 \times -6 \div -3$	**e** $4 \times (-3 - 5)$	**f** $-8 - (7 + 5)$
g $(-3 - 12) \div -3$		

7 Add brackets to make each of these calculations true.

    a $9 \times -2 + 1 = -9$ **b** $-4 + -6 \div -2 = -1$ **c** $5 - 7 + 2 - 4 = -8$

8 **a** $20 \div (-2)^2$ **b** $(-5 - 3)^2$ **c** $5 + 4 \times (-3)^2$ **d** $40 - (4 - 9)^2$

Practice

1B HCF and LCM

1 **a** Write the first 10 multiples of these numbers.
    i 3   **ii** 6   **iii** 8   **iv** 15
   b Use your answers to **a** to find the LCM of each of these pairs of numbers.
    i 3 and 8   **ii** 6 and 15   **iii** 6 and 8   **iv** 8 and 15

2 **a** Write out all the factors of these numbers.
    i 12   **ii** 18   **iii** 20   **iv** 30
   b Use your answers to **a** to find the HCF of each of these pairs of numbers.
    i 12 and 18   **ii** 12 and 20   **iii** 20 and 30   **iv** 18 and 20

3 Find the LCM.

    a 4 and 8   **b** 6 and 10   **c** 7 and 8   **d** 9 and 12

4 Find the HCF.

    a 14 and 35   **b** 8 and 20   **c** 12 and 30   **d** 15 and 24

5 **a** Two numbers have a LCM of 20 and a HCF of 2. What are they?
   b Two numbers have a LCM of 36 and a HCF of 6. What are they?
   c Two numbers have a LCM of 30 and a HCF of 1. What could they be?

6 The LCM of two numbers is twice their HCF. What can you say about the two numbers? Investigate.

Practice

1C Powers and roots

1 Find the cubes of these numbers. Do *not* use your calculator.

    a 5   **b** 11   **c** 12

2 Calculate these powers. Do *not* use your calculator.

    a 6^4 **b** $(-2)^3$ **c** $(-4)^4$ **d** $(-10)^5$

3 Use your calculator to find these.

    a 24^2 **b** 15^3 **c** 25^3

    d 6.6^2 **e** 4.2^3 **f** 7.3^3

4 Find four cube numbers that are also multiples of 3.

5 Use your calculator to find these.

    a 2^6 **b** 4^5 **c** 8^4 **d** 6^6 **e** 3^8

6 Write down the values of these roots.

    a $\sqrt{16}$ **b** $\sqrt{81}$ **c** $\sqrt[3]{64}$ **d** $\sqrt[3]{343}$

7 For each equation, find two values of x that make the equation true.

    a $x^2 = 49$ **b** $x^2 = 100$ **c** $x^2 = 225$ **d** $x^2 = 1.44$

8 **a** Use your calculator to find these.

        i 0.4^2 **ii** 0.5^2

    b What do you think will be the answer to 0.6^2? Try it and see.

    c What do you think will be the answer to 0.3^2? Try it and see.

    d Copy and complete this table.

Number	0.1	0.2	0.3	0.4	0.5	0.6	0.7	0.8	0.9	1
Square										

9 **a** Choose any number between 0 and 1, for example, 0.4.

    b Calculate increasing powers of your number, for example, 0.4^2, 0.4^3, 0.4^4, etc.

    c What do you notice about the sizes of your answers as the power increases? Explain why this happens.

10 **a** Find two square numbers that have exactly one cube number between them. Write all three numbers down in order.

    b Find two cube numbers that have exactly two square numbers between them. Write all four numbers down in order.

11 **a** Use your calculator to find these.

        i 0.5^3 **ii** 0.6^3

    Look at your table of cubes.

    b What do you think will be the answer to 0.7^3? Try it and see.

    c What do you think will be the answer to 0.4^3? Try it and see.

    d What do you think will be the answer to 0.2^3? Try it and see.

    e Copy and complete this table.

Number	0.1	0.2	0.3	0.4	0.5	0.6	0.7	0.8	0.9	1
Cube										

1D Prime factors

1 Calculate these products of prime factors.

   a $2 \times 5 \times 5$
   b $2^3 \times 3$
   c $2 \times 3 \times 7^2$

2 Use a prime factor tree to write each of these numbers as a product of its prime factors.

   a 6
   b 18
   c 32
   d 70
   e 36

3 Use the division method to write each of these numbers as a product of its prime factors.

   a 14
   b 45
   c 96
   d 130
   e 200

4 Calculate the HCF and LCM of each pair of numbers. Use the diagrams to help.

a 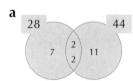 **b**  **c**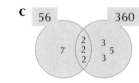

5 $40 = 2 \times 2 \times 2 \times 5$ and $84 = 2 \times 2 \times 3 \times 7$

Make a diagram as in Question 4. Use the diagram to find the HCF and LCM of 40 and 84.

6 Use prime factors to calculate the HCF and LCM of each of these pairs of numbers.

   a 60 and 135
   b 48 and 80
   c 70 and 126

1E Sequences 1

1 Follow these instructions to generate sequences.

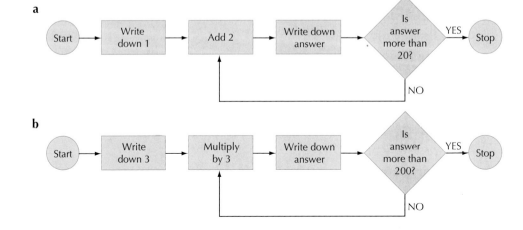

2 **a** Describe how these sequences are generated.
   i 5, 8, 11, 14, 17 …   **ii** 2, 10, 50, 250 …
   b Find the next two numbers in each sequence.

3 You are given a start number and a multiplier. Write down the first five terms of the sequence.

   a Start 1, multiplier 20   **b** Start 64, multiplier $\frac{1}{4}$
   c Start –3, multiplier –2

4 These patterns of dots generate sequences of numbers.

   a

   4   7   10   13

   b

   2   5   9   14

   i Draw the next two patterns of dots.
   ii Write down the next four numbers in the sequence.

5 Draw a flow diagram to generate the sequence that starts with 10 and uses the term-to-term rule 'add 5'.

Practice

1F Sequences 2

1 For each of these arithmetic sequences, write down the first term a and the constant difference d.

   a 6, 8, 10, 12, 14, …   **b** 30, 35, 40, 45, …   **c** 100, 96, 92, 88, …

2 Given first term a and the constant difference d, write down the first six terms of each sequence.

   a $a = 7, d = 6$ **b** $a = 2, d = 2.5$ **c** $a = 8, d = -5$

3 This diagram can be used to generate sequences.

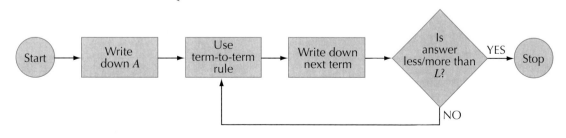

Write down the sequences generated using these values of A and L.

	A	Term-to-term rule	L
a	2	Treble	100
b	20	Subtract 3	−10
c	1	Add 2, 4, 6, 8 etc.	30
d	1	Multiply by 4, subtract 10	−600
e	500	Divide by 10	0.001

(4) **i** The nth term of each sequence is given below. Write down the first five terms.

ii What is the constant difference in each sequence?

   a $3n - 1$ **b** $2n + 5$ **c** $5n - 3$ **d** $10n + 10$

(5) Find the nth term of each sequence.

   a 7, 13, 19, 25, 31, … **b** 5, 8, 11, 14, 17, … **c** 8, 14, 20, 26, 32, …
   d 1, 4, 7, 10, 13, …   **e** 16, 23, 30, 37, 44, …

(6) Write down a first term A and a term-to-term rule that you can use in the flow diagram in Question 3 so that:

   a each term of the sequence is a negative odd number.
   b the first three terms are positive, and the remainder are negative.
   c the first three terms are whole numbers, and the remainder are decimal fractions.

Practice

1G Solving problems

(1) Paving slabs 1 metre square are used for borders around L-shaped ponds.

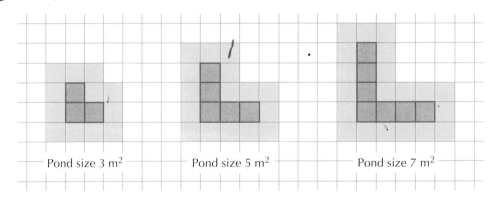

Pond size 3 m²   Pond size 5 m²   Pond size 7 m²

   a How many slabs would fit around a pond of size 9 square metres?
   b Write a rule to show the number of slabs needed to make a border around L-shaped ponds.

2 Write a rule to show the number of slabs needed to make a border around ponds of this shape.

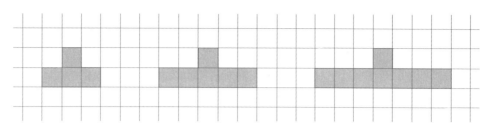

FM **3** Martha opened a post office account on Monday. She deposited £8.
On Friday, she withdrew half the money in her account.
Every Monday, she deposited £4 more than the previous Monday.
Every Friday, she withdrew half the money in her account.

  a Make a table of the amount of money in her account on the first six Saturdays.
  b Write a rule for the amount of money in her account on a Saturday.

 CHAPTER **2**

Geometry and Measures **1**

Practice

2A Alternate and corresponding angles

1 Copy and complete these sentences.

  a *f* and ___ are alternate angles.
  b *c* and ___ are corresponding angles.
  c ___ and *a* are corresponding angles.
  d *c* and *e* are _____ angles
  e *d* and *h* are _____ angles.

 2 **a** Write down three pairs of alternate angles.
  b Write down three pairs of corresponding angles.

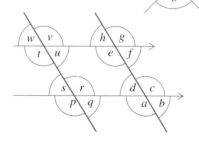

3 Copy each of these diagrams. Calculate the sizes of all the angles and mark them on your diagram.

a

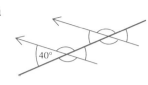

b

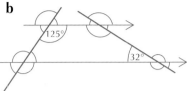

c

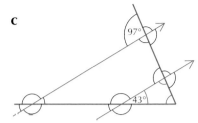

d

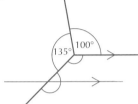

e

f

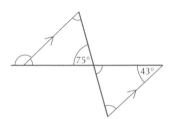

g

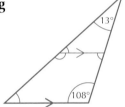

h

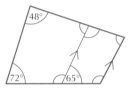

Practice

2B Interior and exterior angles of polygons

1 **a** Find the sum of the interior angles of a nonagon.
b Find the size of an interior angle of a *regular* nonagon.
c Find the size of an exterior angle of a *regular* nonagon.

2 Repeat Question 1, this time for a regular heptagon. Give your answers to **b** and **c** correct to 1 decimal place.

3 Calculate the size of the unknown interior angle.

a

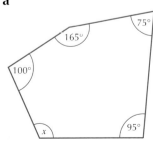

b

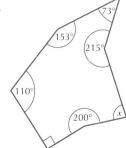

c

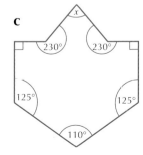

4 Calculate the size of the unknown exterior angle.

a

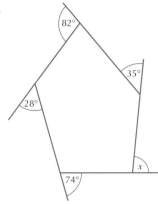

b

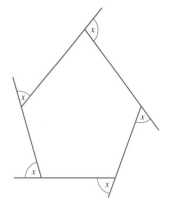

5 How many sides do these regular polygons have?

a

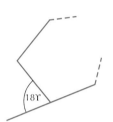

b

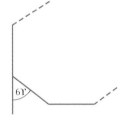

c

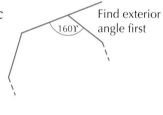

Find exterior angle first

6 For each interior angle of these regular polygons, do the following.

a

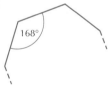

b

i Calculate the size of the exterior angle.
ii Find the number of sides the polygon has.
iii Calculate the sum of its interior angles.

Practice

2C Geometric proof

1 Write a proof to show that opposite angles are equal, that is, $a = c$ and $b = d$.
(Hint: Use angles on a straight line.)

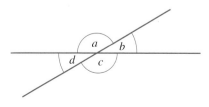

2 ABC is an isosceles triangle.
AD bisects angle A.

Write a proof to show that AD is
perpendicular to BC, that is, $b = c = 90°$.

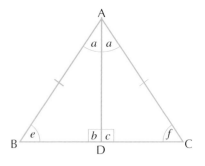

3 Write a proof to show that the opposite angles of a parallelogram are equal.

 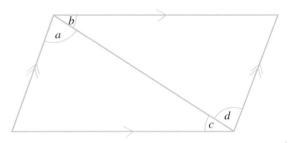

(Hint: Draw a diagonal, as shown on the right.)

 Practice   **2D The geometric properties of quadrilaterals**

1 Copy this table. In each column, write the names of all possible
quadrilaterals that could fit the description.

2 pairs of equal angles	Rotational symmetry of order 4	Exactly 1 line of symmetry	Exactly 2 right angles	Exactly 4 equal sides

(Hint: a quadrilateral could be in more than one column!)

2 A quadrilateral has two pairs of equal sides and three obtuse angles.
What type of quadrilateral is it?

3 A quadrilateral has three equal angles.
What type of quadrilateral could it be?

4 Which quadrilaterals have, or could have, diagonals that intersect at right
angles? Illustrate your answer with drawings.

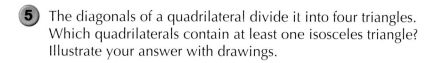

(5) The diagonals of a quadrilateral divide it into four triangles.
Which quadrilaterals contain at least one isosceles triangle?
Illustrate your answer with drawings.

Practice

2E Constructions

(1) **a** Draw a line AB, 13 cm long.
Use a ruler and compasses to bisect the line.
Check the bisection using your ruler.
b Draw another line AB, 13 cm long.
Mark point C on the line, 5 cm from C.
Construct the perpendicular to AB through point C.

(2) Trace this diagram.
Drop a perpendicular from point C to the line AB.

○ C

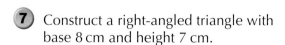

A ——————————————————————————————————— B

(3) Use a protractor to draw a 64° angle. Use a ruler and compasses to bisect
the angle. Check the bisected angles using your protractor.

(4) Repeat Question 3 for the angle 134°.

(5) **a** Draw a line XY, 12 cm long.
b Label the point 5 cm from X as point A.
c Draw a perpendicular to XY through point A.
d By measuring the length of the perpendicular, draw a kite with diagonals
of length 12 cm and 4 cm.

(6) Construct this right-angled triangle.

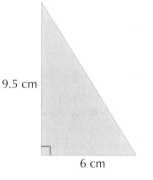

9.5 cm

6 cm

(7) Construct a right-angled triangle with
base 8 cm and height 7 cm.

8 **a** Place your outstretched hand on a sheet of paper. Mark the end of your thumb, middle finger and little finger.
Label the points A, B and C respectively.

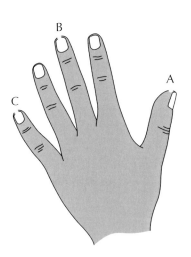
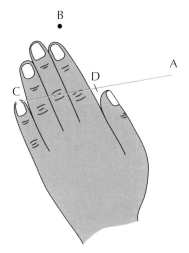

b Draw the line AC. Bisect it using ruler and compasses.
Label the midpoint D.
c Compare the width of your closed hand to the length AD.
d Drop a perpendicular from point B onto AC. Label the intersection E.

CHAPTER **3** Statistics **1**

3A Probability

1 I roll a die 60 times. How many times should I expect to see the following?

a The number 4   **b** An odd number   **c** A number greater than 4

2 A pack of cards is shuffled and one is drawn at random. This happens 100 times.

How many times would you expect to see the following?
a A black card   **b** A diamond   **c** A Jack
d Either a nine or a ten

3 If I toss a fair coin 50 times, how many tails would I expect to see?

4 **a** Write down the eight different results after flipping a 1p coin, a 5p coin and a 10p coin.
   b If you flipped the three coins 100 times, how many times would you expect to see the following?
      i Three heads   **ii** Two tails exactly   **iii** At least one head

5 A coin is flipped 80 times. It lands on heads 37 times.
   Do you think that the coin is biased? Explain your answer.

3B Probability scales

1 Copy and complete this table.

Event	Probability of event occurring (p)	Probability of event *not* occurring ($1 - p$)
A	$\frac{2}{5}$	
B	0.75	
C	$\frac{5}{8}$	
D	0.12	
E	$\frac{19}{20}$	
F	0.875	
G	74%	

2 The probability of an egg having a double yoke is 0.009.
   What is the probability that an egg does *not* have a double yoke?

3 A card is chosen at random from a pack of 52 playing cards. Calculate the probability that it is the following.

   a Not an ace   **b** Not a diamond   **c** Not a picture card
   d Not the six of hearts

4 Joe has a CD stuck on Random in his car CD player. It has 8 tracks of heavy rock, 4 tracks of white rock and 6 tracks of jazz rock.

   a What is the probability that the first track to be played is heavy rock?
   b On a long journey, Joe reckons he would be listening to about 100 tracks. How many of these would he expect to be the following?
      i Not heavy rock   **ii** White rock   **iii** Not jazz rock

5 A weather forecaster estimates the probability of rain to be 35%, black ice 0.82, and a 1 in 10 chance of snow. What is the probability of each of the following?

   a No snow   **b** No black ice   **c** No rain

6 This matchbox cover was dropped many times.

The experimental probabilities of how it could land are shown in this table.

Landing position	Probability
Showing top	0.32
Side	0.23
End	
Showing bottom	0.3

a Calculate the probability of the matchbox cover landing in the following positions.
  i On end
  ii Showing its top or bottom
  iii On its side or end
  iv Not on end
b Comment on the probabilities of the matchbox cover landing on its top or bottom.
c If the box cover was dropped another 1000 times, how many times would you expect it to land showing the bottom?

Practice **3C Mutually exclusive events**

1 This illustration shows diagrams of 8 faces.

Which of these pairs of events are mutually exclusive?
(Hint: 'Left eye' means the eye on the left of the diagram.)

a A smiling face. A sad face.
b Left eye shut. Both eyes shut.
c Wearing a hat. Both eyes open.
d Both eyes open. A sad face.
e Wearing a hat. Right eye shut.
f Smiling with an eye open. Right eye shut.

2 A six-sided die is numbered 1, 2, 3, 3, 4, 6. It is rolled once. Here are some events.

i Number 3
ii Even number
iii Number greater than 3
iv Square number
v Triangle number
vi Multiple of 3
vii Prime number
viii Number 1

a Write down three pairs of events that are mutually exclusive.
b Write down three pairs of events that are *not* mutually exclusive.

3 Two of the dice in Question 2 are rolled together.

   a Make a list of the possible outcomes, for example, (1, 6).
   b What is the probability that the numbers rolled are the following?
     i The same, for example, (3, 3)
     ii One odd, one even, for example, (3, 6)
     iii Both prime, for example, (2, 5)
     iv Both multiples of 3

3D Calculating probabilities

1 The letters of the word SUCCESSOR are written on cards and placed in a bag. One of the cards is withdrawn from the bag.
What is the probability that the letter is the following?

   a An S
   b A vowel
   c One of the last 10 letters of the alphabet
   d A C or an S
   e A consonant
   f A letter of the word ROSE

2 The names of two children, KIM and FRANZ, are written on cards and placed in a bag. These coins are placed in the same bag: 1p, 2p, 5p, 10p, 20p, 50p, £1. A card and a coin are taken from the bag at random. The coin is given to the named person.

   a Make a table to show the possible outcomes.
   b Calculate these probabilities.
     i Kim receives 20p.
     ii Franz receives less than 10p.
     iii One of the children receives 10p.
     iv Kim does not receive £1.
     v Neither child receives more than 20p.

3 A taxi firm owns a red and a green taxicab. The red taxi can carry up to six passengers. The green taxi can carry up to five passengers.

   a Copy and complete this table showing the total number of passengers being carried at any one time.

		Red taxi			
		1	**2**	**3**	
	1	2			
Green taxi	**2**				
	3			6	

   b What is the probability, at any one time, of these numbers of passengers being carried?

     i 7   **ii** 2   **iii** 12
     iv Less than 5   **v** An odd number   **vi** 2 or 7

3E Experimental probability

1 The number of days it rained over different periods are recorded below.

Recording period	Number of days of rain	Experimental probability
30	12	
60	33	
100	42	
200	90	
500	235	

a Copy and complete the table.
b What is the best estimate of the probability of it raining?
 Explain your answer.
c Estimate the probability of it *not* raining.
d Is there a greater chance of it raining or not raining?

2 Colour ten identical matchsticks as shown here and place them in a bag.

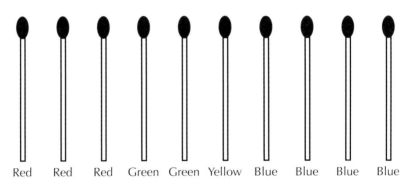

Red   Red   Red   Green   Green   Yellow   Blue   Blue   Blue   Blue

a Withdraw a matchstick from the bag and record its colour in the tally chart below. Do this 10 times.

Colour	Tally	Frequency	Experimental probability	Theoretical probability
Red				
Green				
Yellow				
Blue				

b Calculate the experimental probabilities.
 Write your answers as decimals.
c Calculate the theoretical probabilities.
 Write your answers as decimals.
d Compare the experimental and theoretical probabilities.
e How could you make sure the experimental and theoretical
 probabilities are closer?
f Repeat the experiment so that the experimental and theoretical
 probabilities are closer.

CHAPTER 4 Number 2

4A Fractions and decimals

1 Write these decimals as fractions with a denominator of 10, 100 or 1000. Then cancel to their simplest form.

    a 0.6   **b** 0.36   **c** 0.65   **d** 0.255   **e** 0.025

2 *Without* using a calculator, convert these fractions to decimals.

    a $\frac{13}{50}$ **b** $\frac{17}{25}$ **c** $\frac{4}{5}$ **d** $\frac{3}{20}$

3 Which of these fractions have recurring decimals? Try to answer the questions without using a calculator. Then check to see if you are correct.

    a $\frac{3}{8}$ **b** $\frac{2}{7}$ **c** $\frac{9}{25}$ **d** $\frac{13}{40}$ **e** $\frac{5}{6}$

    f $\frac{19}{50}$ **g** $\frac{4}{9}$ **h** $\frac{3}{14}$ **i** $\frac{15}{16}$ **j** $\frac{16}{21}$

4 Find the larger of each pair of fractions.

    a $\frac{3}{25}$ and $\frac{1}{7}$ **b** $\frac{11}{18}$ and $\frac{9}{16}$ **c** $\frac{7}{11}$ and $\frac{29}{40}$

5 Write these fractions in increasing order of size.

    $\frac{5}{7}, \frac{13}{16}, \frac{33}{50}, \frac{22}{27}$

6 **a** Calculate $\frac{2}{9}$ using your calculator. Describe the display on your calculator.

    b Calculate $\frac{24}{99}$ using your calculator. Describe the display on your calculator.

    c Find fractions that give these calculator displays.

        i 0.888888888   **ii** 0.737373737

        iii 0.657657657   **iv** 0.123412341

        v 0.044444444   **vi** 0.036363636

        vii 0.020202020   **viii** 0.002828282

4B Adding and subtracting fractions

1 Find the lowest common multiple of each of these pairs of numbers.

    a 2, 5   **b** 6, 8   **c** 2, 6   **d** 10, 15   **e** 9, 12

For Questions 2–4 convert to equivalent fractions with a common denominator. Then work out the answer. Cancel your answers and write them as mixed numbers if necessary.

2 a $\frac{2}{5} + \frac{1}{2}$ b $\frac{5}{8} + \frac{1}{12}$ c $\frac{3}{5} - \frac{1}{2}$ d $\frac{5}{9} - \frac{1}{6}$ e $\frac{3}{7} + \frac{2}{3}$

   f $\frac{5}{12} + \frac{1}{8}$ g $\frac{3}{5} + \frac{1}{2} + \frac{7}{10}$ h $\frac{7}{9} - \frac{1}{2}$ i $\frac{11}{15} - \frac{2}{5}$ j $\frac{2}{3} + \frac{5}{6} - \frac{5}{12}$

3 a $2\frac{1}{5} - \frac{7}{10}$ b $1\frac{3}{4} + \frac{5}{6}$ c $3\frac{1}{8} - 2\frac{2}{3}$ d $4\frac{1}{10} - 2\frac{1}{4}$

4 a $\frac{11}{18} - \frac{7}{24}$ b $\frac{19}{40} + \frac{43}{60}$ c $2\frac{3}{25} + 4\frac{7}{10}$ d $8\frac{3}{16} - 5\frac{11}{12}$

Practice

4C Multiplying and dividing fractions

1 Use any method to work out the following.

   a $\frac{1}{4}$ of 28   b $\frac{3}{5}$ of 30   c $\frac{1}{3}$ of 36   d $\frac{3}{4}$ of 24

2 Calculate these.

   a $\frac{4}{9}$ of 36 kg   b $\frac{5}{12}$ of 600 ml   c $\frac{2}{7}$ of 98 cm

   d $\frac{1}{4}$ of 52 km   e $\frac{3}{7}$ of 42 cm   f $\frac{7}{10}$ of 50 grams

   g $\frac{3}{5}$ of £80   h $\frac{7}{8}$ of 640 litres

3 Work out these. Cancel your answers and write them as mixed numbers if necessary.

   a $4 \square \frac{3}{7}$ b $9 \square \frac{5}{6}$ c $7 \square \frac{3}{10}$ d $8 \square \frac{3}{4}$

4 Work out each of following.

   a $\frac{2}{5} \div 3$ b $\frac{8}{9} \div 6$ c $\frac{2}{3} \div 6$ d $\frac{3}{7} \div 4$

5 a $3\frac{3}{4}$ cakes were shared equally among 12 people. What fraction of a cake did each person receive?

   b Kailash measured the length and breadth of this picture frame he made.

     i What is the total length of frame material used?

     ii Kailash has 50 inches of frame material to make an identical frame.
How much frame material will be left over?

     iii He bought a 97-inch strip of frame material. How many of the above frames could he make from it?
Show your working.

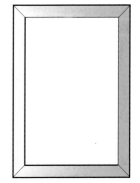

4D Percentages

1 Without using a calculator, express the following.

    a 19 as a percentage of 25   **b** 13 as a percentage of 20

    c 96 as a percentage of 400   **d** 3 as a percentage of 50

2 Use a calculator to express the following, correct to the nearest per cent.

    a 14 as a percentage of 23   **b** 81 as a percentage of 120

    c 6 as a percentage of 65   **d** 3200 as a percentage of 7000

3 Milton drinks 35 cl of an 80-cl bottle of orange.

    a What percentage did he drink?   **b** What percentage remains?

4 A company employs 280 workers. At the last general election, 130 voted Labour, 83 voted Liberal Democrat and the remainder voted Conservative. Calculate the percentage vote for each party.

5 Three people contributed £23, £18.50 and £25.60 toward their total restaurant bill. What percentage of the total bill did each person contribute?

6 **a** Stuart took a maths test lasting 80 minutes. He spent 23 minutes on the Algebra section, 33 minutes on the Number section and the remaining time on the Geometry section.
What percentage of the time did he spend on each section?

    b Marcel spent the same amount of time on each section as Stuart. But he also took an Extension section lasting 30 minutes. What percentage of the time did he spend on each section?

4E Percentage increase and decrease

Do not use a calculator for Questions 1–4.

1 Calculate each of the following.

    a 70% of 300 cm   **b** 5% of £30

    c 35% of 4000 kg   **d** 85% of 8 inches

2 **a** Increase $40 by 20%.   **b** Decrease 200 kg by 5%.

    c Decrease 7p by 60%.   **d** Increase 2000 m by 95%.

    e Increase 230 ml by 45%.   **f** Decrease £7.60 by 85%.

3 **a** Before a typing course, Leon could type 60 words per minute. The typing course increased his speed by 35%. What was his speed after the course?

    b Leon injured his hand, reducing his speed by 20%. What is his speed now?

4 Three people each withdrew a certain percentage of their bank balance from their bank account.

> John: 29% of £300
> Hans: 75% of £640
> Will: 15% of £280

Which person has the largest remaining bank balance?

Use a calculator for Questions 5–9.

5 Calculate the following.

   a 34% of $45   **b** 89% of £33 290

   c $17\frac{1}{2}$% of £90   **d** $5\frac{1}{2}$% of 9 000 000 tons

   e 52.9% of 700 litres   **f** 2.1% of £5300

6 **a** Increase 254 cm² by 18%.   **b** Decrease £15.23 by 84%.
   c Increase 7000 g by 47.2%.   **d** Increase 0.74 litres by 31%.
   e Decrease £17 400 by 97%.   **f** Decrease 900 cm by 2.8%.

7 A coat costs £72 before VAT of 17.5% is added.
What is the price including VAT?

8 Marvin's average score on the computer game Space Attack was 256.

   a After buying a new gamepad, his average score increased by 28%.
What was his new average score, to the nearest whole number?
   b Marvin went on holiday. When he returned, his average score had decreased by 15%.
What was his new average score, to the nearest whole number?

9 'Zipping' a computer file reduces its size by a certain percentage.
Find the size of each file after zipping.

   a 500 kB file reduced by 17%   **b** 740 kB file reduced by 43%
   c 655 kB file reduced by 23%   **d** 1200 kB file reduced by 69.2%

Practice

4F Real-life problems

1 Marcia has an annual salary of £24 000. Her tax allowance is £4800.
She pays tax at 20% on the remainder.

   a How much tax does she pay altogether?

   b How much of her annual salary does she receive?

   c What is her monthly income?

   d She pays 3% of the salary she received toward a pension scheme.
How much is this?

   e In 2007, she received a 7% increase in her annual salary. Her tax allowance remained the same. Recalculate the answers to parts **a** to **d**.

2 Mrs Walker left £45 750 in her will. The money was to be divided between her three children as follows.

Derek 37%   Maria 29%   Jason 34%

The solicitors deducted 6% for expenses. The remainder was divided among the children. How much did each child receive?

 3 Three dealers offer the following repayment options for a car with a marked price of £12 600.

Trustworthy Cars   20% deposit followed by 12 monthly payments of £900
Bargain Autos   114% of the marked price spread over 18 months
Future Car Sales   7% of the marked price for each of the first 6 months followed by 6 monthly payments of £1300

Which deal works out the cheapest overall?

 4 Amanda deposited £3000 in her bank. She kept the money in the bank for 5 years. The bank paid interest of 4% per annum. Calculate the amount of interest Amanda received after 5 years if she does the following.

a Spent the interest at the end of each year

b Added the interest to her bank account at the end of each year

(Hint: Work out the amount in her account at the end of Year 1, then Year 2 and so on.)

 **Algebra 2**

Practice

5A Algebraic shorthand

1 Write each of these expressions using algebraic shorthand.

a $x \div 5$ **b** $3 \times b$ **c** $4 \times m \times n$
d $p \div q$ **e** $2 \times (x - 1)$ **f** $5 \times m + 3$
g $c \div (d + 2)$ **h** $5 \times t \div 4$ **i** $(a + b) \times (m - n)$

2 Simplify these expressions.

a $6n \times t$ **b** $5k \times 3m$ **c** $h \times 5g$ **d** $2u \times 3t \times 4v$

3 Simplify these expressions.

a $4x \div 4$ **b** $\dfrac{9y}{9}$ **c** $\dfrac{8a}{2}$ **d** $\dfrac{7b}{b}$ **e** $\dfrac{20i}{5i}$

4 Solve each of these equations.
In your solution, write each equation on a separate line.

a $5x + 7 = 22$ **b** $2p - 9 = 7$ **c** $10m + 15 = 125$
d $12c - 13 = 71$ **e** $4k + 8 = 19$ **f** $8f - 25 = 98$

5 Which of these statements are incorrect?
If you are unsure, check by substituting a number for the letter.

a $6 \div d$ is the same as $\dfrac{d}{6}$

b $4(a + 2)$ is the same as $2(a + 4)$

c $\dfrac{10m}{5}$ is the same as $2m$

d $\dfrac{12}{3n}$ is the same as $4n$

e $(a - 2) \div 2$ is the same as $2 \times (a + 2)$

f $6 \times d \div 2$ is the same as $\dfrac{6d}{2}$

6 Which of these statements are incorrect?

a $5 \times p = 12$ is the same as $12 = 5p$
b $5m + 2 = 17$ is the same as $17 = 5m + 2$
c $9 = 2j - 4$ is the same as $2j + 4 = 9$
d $n + 3 = 12$ is the same as $12 = 3 + n$
e $2t + 3 = 9$ is the same as $3 = 2t + 9$
f $7 - 3p = 10$ is the same as $10 = 3p - 7$

7 Write down the equivalent expressions, for example, $5d = 5 \times d = d5$.

$st + 3$

$3 - st$

$3s + t$

$3 + t \times s$

$3 \div s + t$ $3 + st$

$3 \times s \times t$

$ts - 3$ $s \times 3 + t$

$t + s3$ $t \times 3 + s$

$\dfrac{s}{3} + t$ $s3t$

$s + 3t$

$t + \dfrac{3}{s}$ $t \times s + 3$

Practice

5B Like terms

1 Make a list of the terms in each of these.

    a $y + 2x - 3$ **b** $4x = 3 + 2x$ **c** $\frac{4}{t} - u$ **d** $3x^2 = 14x + 2$

For Questions 2–5, simplify the expressions.

2
  a $4i + 7i$ **b** $7r - 2r$ **c** $-u - u$
  d $3h + 2h - h$ **e** $6y - 8y$ **f** $-3m + 5m$
  g $4t - 3t - 6t$ **h** $4n - 3n$ **i** $5zt + 4zt$
  j $-ab - ab$ **k** $2ad + 7ad - 10ad$ **l** $3f^2 - f^2$

3
  a $6k + 4k + 3l$ **b** $7h + 3i - 2i$ **c** $10y + 2x + 3y$
  d $4p + 2p - 1$ **e** $4d - 7d + 2e$ **f** $7t - 2u - 4t$
  g $9w + 3x - 10w$ **h** $3fg + 7f + 2fg$ **i** $s^2 - t^2 + 3s^2$

4
  a $4q + 3q + 6i + i$ **b** $8z - 3z + 4b - 2b$
  c $9u + 3v + 2u + 4v$ **d** $7j + 5k - 3k + 2j$
  e $3m - 5m - 4n + 7n$ **f** $4d + 6e - 8e + 2d$
  g $-3c + 4d + 8c - 5d$ **h** $10r - 4s - 7s - 8r$
  i $g - h - 2g + 2h$

5
  a $2m^2 + m^2 + 4m - m$ **b** $5h^2 + 3h + 7h - 3h^2$
  c $2p^2 + 6p + 4p^2 - 4p$ **d** $8d^2 - 3d - 5d^2 - 2d$

Practice

5C Expanding brackets and factorising

1 Expand the brackets.

  a $2(f + k)$ **b** $s(t - 4)$ **c** $4(2a + 3)$ **d** $m(n + 3r)$
  e $3(3w - 2s)$ **f** $a(2b - 3)$ **g** $5(s + t - u)$ **h** $d(2 - 3r + 4g)$

2 Expand the brackets.

  a $-(p + q)$ **b** $-(3r - s)$ **c** $-4(a + b)$ **d** $-2(m - n)$ **e** $-5(2d - 3e)$

For Questions 3–6, expand and simplify the expressions.

3
  a $4(f + g) + 6f$ **b** $2k + 3(3k - s)$ **c** $4x + 2(2y - 3x)$

4
  a $5(p + q) + 2(2p + 3q)$ **b** $4(2i + j) + 3(i - j)$
  c $2(3b - 2a) + 4(b - 2a)$ **d** $5(m - 3n) + 2(3m + 4n)$

5
  a $4t - (2t + 3u)$ **b** $7m - (2m - n)$ **c** $2x - (4y + 3x)$

6
  a $4(h + i) - (2h + 3i)$ **b** $6(2s + t) - (3s - 2t)$
  c $5(2w - v) - 3(4w + 2v)$

7 Factorise these expressions by filling in the boxes.

a $10m - 8 = \boxed{}(5m - 4)$ **b** $12d + 8e = 4(\boxed{} + \boxed{})$

c $st + 4s = \boxed{}(t + 4)$ **d** $5de - 2e = e(\boxed{} - \boxed{})$

e $2n + 6 = \boxed{}(\boxed{} + \boxed{})$ **f** $14sf - 7 = \boxed{}(\boxed{} - \boxed{})$

g $t - ty = \boxed{}(\boxed{} - \boxed{})$ **h** $3mh + 5nh = \boxed{}(\boxed{} + \boxed{})$

8 Factorise these expressions.

a $6f + 9$ **b** $8 - 4s$ **c** $df + de$ **d** $6ab - 5a$

e $12v^2 - 8$ **f** $8w - 2$ **g** $rt - r$ **h** $15st - 10$

Practice   **5D Using algebra and shapes**

1 Write down the perimeter and area of each shape as simply as possible.

a

5

$2x - 1$

b

3

$4n + 1$

c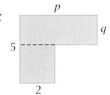

p

q

5

2

d

5

b

2

$a + 1$

e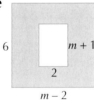

6

$m + 1$

2

$m - 2$

2 a Write down the area of the large parallelogram.
Area of a parallelogram = base × height

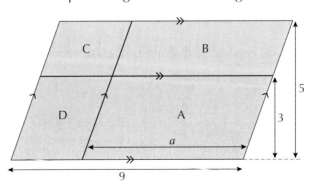

C   B

D   A

a

5

3

9

b Write down the areas of the smaller parallelograms, A to D.
c Show that the sum of the smaller areas is the same as your answer to part **a**.

3 For each question, mark any necessary lengths on your diagram.

    a Sketch a rectangle whose area is abc.
    b Sketch a rectangle whose perimeter is $6x + 4y$.
    c Sketch a triangle whose area is $8x$.
    d Sketch a rectangle whose area is $5x + 10$.

5E Index notation with algebra

1 Write these expressions using index form.

    a $g \times g \times g \times g \times g \times g$ **b** $9k \times k$ **c** $5t \times 3t$
    d $r \times 4r$ **e** $2j \times 2j \times 2j$

2 Write these expressions as briefly as possible.

    a $m + m + m + m + m + m + m + m$
    b $t \times t \times t \times t \times t$ **c** $d + d + d + e \times e \times e$

3 Explain the difference between $6w$ and w^6.

4 Expand the brackets.

    a $v(3 + v)$ **b** $m(3m - 2n)$ **c** $D(3E - 2D)$ **d** $3s(2s + 3t)$

5 Expand and simplify these expressions.

    a $2ab + a(3b + 2)$ **b** $v(2v + 4t) - 2vt$ **c** $9qz - q(3z - 2q)$

6 Expand and simplify these expressions.

    a $r(2r + s) + s(3r + s)$ **b** $p(3p - 2q) + q(4p - 5q)$
    c $m(3m + 2n) - n(2m + 4n)$ **d** $c(3c - 2d) - d(5c - d)$

7 Expand and simplify these expressions.

    a $2g^2 + g(3g + 4)$ **b** $d(2d + 4e) + d(3d - e)$
    c $w(4 - 2w) + w(4w + 3)$

8 Simplify the following.

    a $d^2 \times d^4$ **b** hh^3 **c** $2m \times 5m^4$ **d** $3a^3 \times 4a^2$

    e $\dfrac{u^5}{u}$ **f** $n^7 \div n^2$ **g** $\dfrac{14u^6}{7u^2}$ **h** $20v^8 \div 4v^3$

9 Expand and simplify the following.

   a $d(d^2 + 2d) + d(3d + 4)$ **b** $4c(c^2 - 3) + 2c^2(c + 5)$

10 Factorise each of the following.

   a $r^2 + 4r$ **b** $5s^2 - 2s$ **c** $a^2b + 3a$ **d** $a^3 - a$ **e** $4w - 5w^3$

CHAPTER 6 Geometry and Measures **2**

Practice

6A The circle

1 **a** Name the lines on this diagram.

   i AB   **ii** AC

   iii DE   **iv** CF

   v CG

b Name these shapes.

   i ABCA   **ii** ACFA

   iii ABCFA   **iv** GABCG

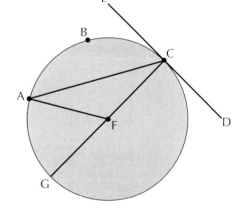

2 **a** Measure the diameter of a bicycle wheel to the nearest centimetre.

b Measure the circumference of the wheel by rolling it along the ground for one turn.

c Use a calculator to divide your answer to **b** by your answer to **a**.

d The answer should be a little more than 3. How much more?

Practice

1 Use the approximation $\pi \approx 3.14$ to calculate the circumference of each wheel. Write your answers correct to the nearest centimetre.

a
60 cm

b
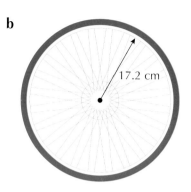
17.2 cm

2 Use the π button of your calculator to calculate the circumference of each coaster. Write your answers correct to 1 decimal place.

a
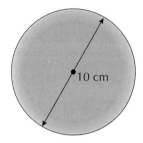
10 cm

b
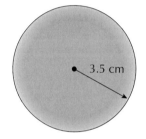
3.5 cm

3 Use the approximation $\pi \approx \frac{22}{7}$ to calculate the circumference of each coin. Do not use a calculator.

a

TWO POUNDS
14 mm
1998

b
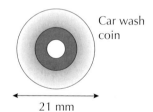
Car wash coin
21 mm

4 Calculate the total length of the lines in this crop circle. It has two semicircles, a circle and straight lines. Write your answers correct to the nearest centimetre.

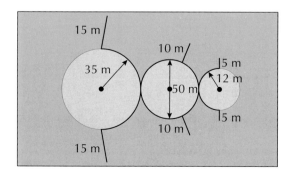
15 m
10 m
35 m
5 m
12 m
50 m
10 m
5 m
15 m

6

7

Use the approximation π ≈ 3.14 or the π button of your calculator for these questions.

1 Calculate the area of the lid of each tin. Write your answers using a suitable degree of accuracy.

a

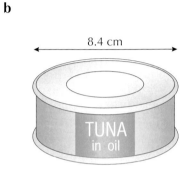

38 mm

b

8.4 cm

c

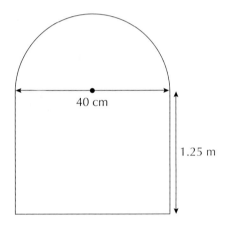

13 cm

2 Calculate the total area of this arched window.
Write your answer correct to the nearest square centimetre.

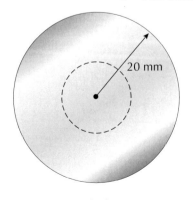

40 cm

1.25 m

3 This diagram shows how a washer is made.

  a Calculate the area of the blank, correct to 1 decimal place.
  b Calculate the area of the hole, correct to 1 decimal place.
  c Calculate the area of the finished washer, correct to the nearest mm².

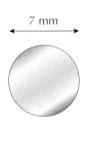

20 mm

7 mm

Blank   Centre removed   Finished washer

1 Convert these quantities.

    a $0.07\,m^2$ to cm^2 **b** $8400\,cm^3$ to l   **c** $21\,000\,000\,cm^3$ to m^3

    d $4.8\,ml$ to mm^3 **e** $5340\,mm^2$ to cm^2 **f** $5200\,l$ to m^3

2   **i**   Calculate the surface area of each prism.

    ii   Calculate the volume of each prism.

a

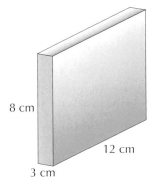

b

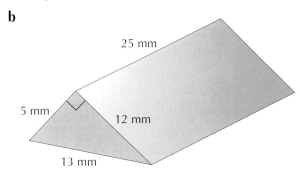

3 Convert your answers to Question **2b** to cm^2 and cm^3.

4 The cross-section of this plastic bench has an area of $620\,cm^2$. Calculate its volume in **a** cm^3 and **b** m^3.

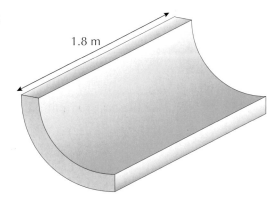

5 The diagram shows a petrol tank in the shape of a prism.

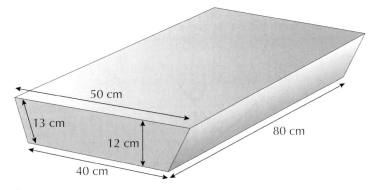

    a   Calculate the area of the cross-section.

    b   Calculate the volume of the tank.

    c   Calculate the capacity of the tank in litres.

    d   Calculate the surface area of the tank in m^2.

5

1 Convert each of these quantities to the units shown in brackets.

   a 25 st (lb)   **b** 7 miles (yd)   **c** 9 ft 7 in (in)   **d** $3\frac{1}{2}$ ton (lb)

2 Convert each of these quantities to the units shown in brackets.

   a 83 oz (lb and oz)   **b** 532 ft (yd and ft)
   c 75 pt (gall and pt)   **d** 125 in (ft and in)

3 **a** How many stones are in a ton?
   b How many feet are in a mile?

4 Convert each Imperial quantity to the approximate metric quantity shown in brackets. Use 1 oz ≈ 30 g and 1 lb ≈ 450 g.

   a 11 oz (g)   **b** 270 miles (km)
   c 100 in (m)   **d** $7\frac{1}{4}$ lb (kg)

5 Convert each metric quantity to the approximate Imperial quantity shown in brackets.

   a 72 km (miles)   **b** 420 g (oz)
   c 81 l (gall and pt) **d** 720 g (lb and oz)

FM **6** Calculate the approximate length of this tape measure in yards, feet and inches.

Use the approximation 1 in ≈ 2.5 cm.

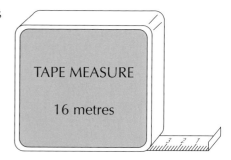

TAPE MEASURE

16 metres

FM **7** A small Dutch beer glass holds 0.25 l of beer. How many pints does it hold?

FM **8** Convert this recipe to Imperial units.

Biscuits

750 g   plain flour
360 g   butter
315 g   sugar
75 g   rice flour
15 g   salt

FM **9** You know that 1 in ≈ 2.5 cm and 1 oz ≈ 30 g.
Find a close metric approximation (shown in brackets) for each of these.

   a 1 yd (cm)   **b** 1 lb (g)   **c** 1 mile (m)

Practice

7A Linear functions

1 a Copy this mapping diagram.

```
-4      -3      -2      -1       0       1       2       3       4
|-------|-------|-------|-------|-------|-------|-------|-------|      x

                                                                  x - 3
-4      -3      -2      -1       0       1       2       3       4
```

b For the function $x \to x - 3$, map the integer values from $x = -1$ to $x = 4$.
c Map these integer values.
  i $x = 3.5$ **ii** $x = 0.5$
  iii $x = 1.5$ **iv** $x = -0.5$

2 a For each of these functions, draw a mapping diagram using two lines from $x = -5$ to $x = 10$. Label your diagrams.
  i $x \to x + 3$ **ii** $x \to x - 1$ **iii** $x \to 2x$
  iv $x \to 2x + 4$ **v** $x \to 2x - 2$ **vi** $x \to 3x + 3$
b Map these integer values.
  i $x = 2.5$ **ii** $x = 0.5$
  iii $x = -1.5$ **iv** $x = -0.5$

3 Draw a mapping diagram for the function $x \to -2x$ using two number lines from $x = -6$ to $x = 6$.

Practice

7B Finding functions from inputs and outputs

For Questions 1–4, find the function that maps the inputs to the outputs.

1 a $\{1, 2, 3, 4, 5\} \to \{7, 8, 9, 10, 11\}$
b $\{3, 4, 5, 6, 7\} \to \{6, 8, 10, 12, 14\}$
c $\{0, 1, 2, 3, 4\} \to \{-2, -1, 0, 1, 2\}$
d $\{1, 2, 3, 4, 5\} \to \{3, 5, 7, 9, 11\}$
e $\{2, 3, 4, 5, 6\} \to \{5, 8, 11, 14, 17\}$
f $\{-2, -1, 0, 1, 2\} \to \{1, 3, 5, 7, 9\}$
g $\{-1, 0, 1, 2, 3\} \to \{-6, -2, 2, 6, 10\}$

2 a $\{3, 4, 5, 8, 9\} \to \{8, 11, 14, 23, 26\}$
b $\{2, 4, 5, 6, 8\} \to \{8, 14, 17, 20, 26\}$
c $\{0, 4, 6, 7, 8\} \to \{-3, 17, 27, 32, 37\}$
d $\{1, 3, 5, 7, 9\} \to \{4, 12, 20, 28, 36\}$
e $\{3, 6, 9, 12, 15\} \to \{3, 9, 15, 21, 27\}$

3 **a** $\{3, 1, 2, 5, 4\} \rightarrow \{12, 4, 8, 20, 16\}$
   b $\{6, 0, 3, 2, 1\} \rightarrow \{13, 1, 7, 5, 3\}$
   c $\{4, 2, 5, 3, 6\} \rightarrow \{22, 16, 25, 19, 28\}$
   (Hint: Rearrange the inputs and outputs in sequence first.)

4 **a** $\{1, 2\} \rightarrow \{5, 7\}$ **b** $\{2, 4\} \rightarrow \{4, 10\}$ **c** $\{0, 3\} \rightarrow \{1, 16\}$

5 For each function, do the following.
   i Find the outputs for the inputs $\{1, 2, 3, 4, 5\}$.
   ii Find the inverse of the function.
   iii Check that the inverse function maps the outputs of **i** to the
      inputs given.
   a $x \rightarrow x + 5$ **b** $x \rightarrow 6x$ **c** $x \rightarrow 2x - 3$ **d** $x \rightarrow 4x + 5$

Practice

7C Graphs of functions

1 **a** Copy and complete this table for the function $y = 4x - 3$.

x	0	1	2	3	4	5
$y = 4x - 3$						

   b Draw a grid with its x-axis from 0 to 5 and y-axis from −5 to 20.
   c Draw the graph of the function $y = 4x - 3$.

2 **a** Copy and complete this table.

x	−2	−1	0	1	2	3
$y = x + 2$						
$y = 2x + 2$						
$y = 3x + 2$						
$y = 4x + 2$						

   b Draw a grid with its x-axis from −2 to 3 and y-axis from −10 to 15.
   c Draw the graph of each function in the table.
   d What is the same about the lines?
   e What is different about the lines?
   f Use a dotted line to sketch the graph of $y = 2.5x + 2$.

3 For the function $8y + 6x = 24$, do the following.
   a Find y when $x = 0$.
   b Find x when $y = 0$.
   c Find a third point that lies on the line.
   d Draw the graph of the function.

4 Draw the graph of the function $10y + 4x = 20$ by finding three points that
   lie on the line.

Practice

7D Gradient of a straight line

1. For each of these lines, write down the following.
   i The gradient **ii** Where it cuts the y-axis

a   **b**   **c**   **d**

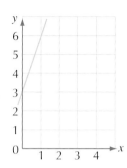
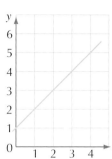
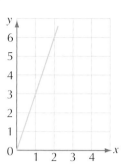
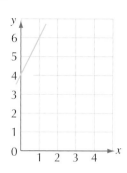

2. Write down the equation of each line in Question 1.

3. **a** Plot the points A(1, 3) and B(3, 5).
   b Work out the gradient of AB.
   c Extend the line to cross the y-axis and give this point.
   d Write down the equation of the line that passes through AB.

4. For each function, do the following.
   i State the gradient.
   ii State the y-intercept.
   iii Draw a graph using squared paper (do not plot any points other than the y-intercept).
   a $y = 3x + 5$ **b** $y = 2x - 2$

Practice

7E Real-life graphs

1. **a** Copy this grid.

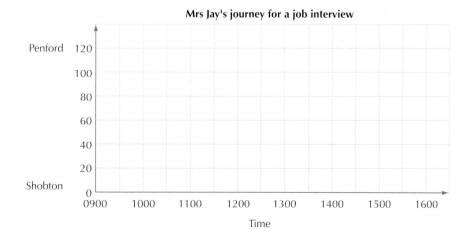

Mrs Jay's journey for a job interview

b Draw on your grid a distance–time graph for Mrs Jay's train journey. Mark the places she visited on the vertical axis.

Mrs Jay had to travel to a job interview at Penford. She caught the 0900 train from Shobton and travelled to Deely 30 miles away in an hour. There she had to wait 30 minutes for a connecting train to Penford. This part of the journey took $1\frac{1}{2}$ hours at a speed of 60 mph. Her job interview at Penford lasted 2 hours. Her return journey lasted $1\frac{1}{2}$ hours.

c Calculate the average speed of the train on her return journey.

 a Draw a grid with these scales.
  - Horizontal axis represents time, from 0 to 6 hours, 1 cm to 30 minutes
  - Vertical axis represents distance from factory, from 0 to 50 miles, 1 cm to 5 miles

b Draw on the grid a distance–time graph for the lorry journey. Mark the places of delivery on the vertical axis.

A car transporter left the factory and took 30 minutes to travel 15 miles to a dealer in Harton. It took half an hour to unload three of the cars. The transporter travelled at 25 mph for another hour and made a delivery at Glimp. This delivery and lunch took an hour. A final delivery was made 30 minutes later at Unwich after a 10-mile drive. This delivery took 30 minutes. The transporter returned to the factory at a speed of 50 mph.

c Calculate the average speed of the transporter:
  i between Glimp and Unwich. **ii** for the whole trip.

 a Draw a grid with these scales.
  - Horizontal axis represents time, from 0 to 30 minutes, 1 cm to 2 minutes
  - Vertical axis represents temperature, from 20°C to 100°C, 1 cm to 10°C

b Draw a graph on the grid to show the temperature of Stan's CupSoup.

Stan's microwave takes 4 minutes to boil a mug of CupSoup at full power. He left the mug in the microwave on full power for 6 minutes. He forgot about the soup for a further 9 minutes, by which time it had cooled to 80°C. So he heated it to boiling point on quarter power. It took him 9 minutes to eat the soup. The temperature of the last spoonful was 50°C.

CHAPTER 8 Number **3**

Practice

8A Powers of 10

Do not use your calculator.

1 Multiply each of these numbers by 10^4.

  a 2.7   **b** 0.05   **c** 38   **d** 0.008

(2) Divide each of these numbers by 10^3.

    a 730   **b** 4   **c** 2.8   **d** 35 842

(3) Calculate these.

    a 7.4×10^3 **b** 13×10^5 **c** $0.87 \div 10^3$
    d $17.4 \div 10^2$ **e** 0.0065×10^3 **f** $19.4 \div 10^4$

(4) Multiply these numbers **i** by 0.01 and **ii** by 0.001.

    a 280   **b** 0.6
    c 1.05   **d** 9851

(5) Divide these numbers **i** by 0.01 and **ii** by 0.001.

    a 4   **b** 800   **c** 0.9   **d** 67.2

(6) Work your way along this chain of calculations for each of these starting numbers.

    a 2000   **b** 7   **c** 0.6

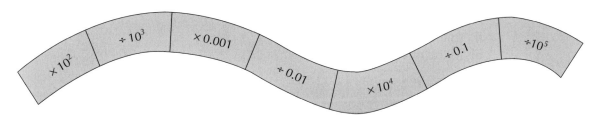

(7) Round these numbers **i** to 1 decimal place and **ii** to 2 decimal places.

    a 8.265   **b** 3.965   **c** 0.047
    d 4.994   **e** 0.095

(8) Work out the following.

    a 14.281×10 **b** 6.3942×10^2 **c** $225.61 \div 10$
    Round your answers to 1 decimal place.

Practice

8B Large numbers

(1) Write these numbers in words.

    a 956 348   **b** 15 230 421
    c 8 002 040   **d** 604 500 002

(2) Write these numbers using figures.

    a Two hundred and six thousand, one hundred and seven
    b Fifty million, thirty-two thousand and eight

4

3 This graph shows the numbers of a company's shares sold every hour during a trading day.

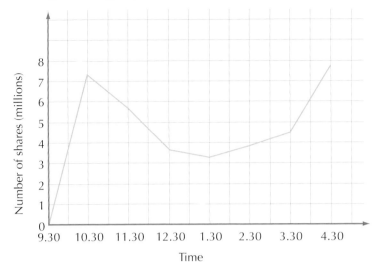

Estimate the number sold each hour. Make a table for your answers.

4 Round these numbers:
i to the nearest ten thousand.
ii to the nearest hundred thousand.
iii to the nearest nearest million.

a 7 247 964   b 1 952 599   c 645 491   d 9 595 902

5 The top six UK airports, in 2006, measured by number of passengers are shown in the table.

1	Heathrow	67 339 000
2	Gatwick	34 080 000
3	Stansted	23 680 000
4	Manchester	22 124 000
5	Luton	9 415 000
6	Birmingham	9 056 000

a Which two of these airports had the same number of passengers to the nearest million?
b How many passengers used Heathrow, Gatwick or Stansted in 2006? Give your answer to the nearest ten million passengers.

6 Use a calculator to find these values.

a 10^{-5} b 10^{-8} c 9×10^{-3}
d 230×10^{-4} e 0.52×10^{-1}

7 Write these standard form numbers out in full.

The first one has been done for you.

a $4.3 \times 10^4 = 43\ 000$ b 5.87×10^8
c 7.01×10^5 d 1.023×10^5

8C Estimations

1. Estimate each value pointed to by the arrows.

   a

   0   4.8

   b

   0   9.2

   c

   −10   10

2. Estimate answers to these.

   a 71% of £589   **b** $\sqrt{80}$ **c** 33×777

   d 7.57^2 **e** $\frac{1}{5}$ of 973   **f** $274 \div 53$

   g $\dfrac{34.9 + 73.8}{62.8 - 38.9}$ **h** 0.745×0.328 **i** $2.502 \div 0.0483$

3. Choose the closest answer in the brackets. Justify your choice.

   a 35.82×73.29 (2350, 280, 2740)
   b 0.834^2 (0.92, 0.06, 0.64)
   c $\sqrt{60}$ (8.3, 6.9, 7.7)

4. Which of these statements is likely to be correct? Explain your answer using estimation and number facts.

   a 34% of 69 = 22
   b 4.55 is halfway between 3.109 and 8.691
   c $\frac{1}{5}$ of 1620 = 323
   d Thirty-nine 0.23 kg bags of salt weigh 18 kg, to the nearest kilogram

8D Working with decimals

Do not use a calculator, apart from Questions 2 and 9. Show your working.

1. Round these numbers **i** to 2 decimal places and **ii** to 3 decimal places.

   a 7.5745   **b** 40.595 99   **c** 0.039 493
   d 8.899 99   **e** 126.039 555

2 Work these out using a calculator. Write your answers correct to 3 decimal places.

 a $1200 \div 2893$ **b** 0.342^2 **c** $\sqrt{5}$ **d** $\dfrac{1}{32.4 - 4.3^2}$

3 Calculate these.

 a $2.06 + 9.77 + 12.3$ **b** $0.87 + 1.79 - 0.94$
 c $78.008 - 23.7 - 19.08$ **d** $9.231 - 2.076 - 1.8$
 e $13 + 91.03 - 2.378 - 18.26 + 33.333$

4 Calculate these. Work in metres.

 a $5\,m - 2.56\,m + 108\,cm$ **b** $0.95\,m + 239\,cm - 1.86\,m$
 c $6\,cm + 0.67\,m - 0.085\,m$ **d** $23.6\,cm + 0.082\,m - 7.41\,cm$

5 **a** Calculate the total volume of juice in these full bottles. Work in litres.

 (Hint: 1 litre = 100 cl = 1000 ml.)

 b All the juice is made into a fruit cocktail. Three cups are drunk.
 If a cup holds 15.3 cl, how much fruit cocktail is left?

6 The skin is the largest organ in the body and weighs 10.886 kg, on average. The four other largest organs are Liver (1.56 kg), Brain (1.408 kg), Lungs (1.09 kg), Heart (0.315 kg).

 a Calculate the total weight of these organs.
 b How much more does the skin weigh compared to the total weight of the other organs?

7 A candle is 32.7 cm tall. It burns down 18.4 mm each day.
 What is its height at the end of 4 days? Work in centimetres.

8 Calculate the area of each of these shapes, giving your answers correct to 3 decimal places.

a

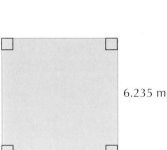

6.235 m

6.235 m

b

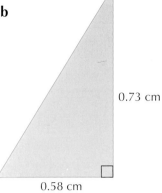

0.73 cm

0.58 cm

c

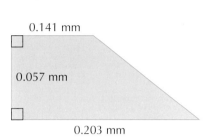

0.141 mm

0.057 mm

0.203 mm

9 **a** Use a calculator to find the value of √2 correct to 8 decimal places.

b Which of these fractions gives the closest approximation to √2? Show your working.

i $\frac{71}{50}$ **ii** $\frac{482}{341}$ **iii** $\frac{3542}{2503}$ **iv** $\frac{10\,600}{7499}$

Practice

8E Efficient calculations

1 Use the bracket and/or memory keys on your calculator to calculate these.

a $19.3 - (32.5 - 24.8)$ **b** $(0.24 + 1.73)^2$ **c** $\dfrac{1}{2.09 - 1.93}$

2 Use the fraction key on your calculator to calculate these.

a $\frac{7}{9} - \frac{1}{6}$ **b** $2\frac{3}{5} + 7\frac{1}{4}$ **c** $9\frac{3}{10} \times 2\frac{1}{2}$

d $1\frac{3}{8} \div \left(\frac{5}{6} - \frac{2}{9}\right)$ **e** $\left(5\frac{1}{4} - 2\frac{1}{7}\right)^2$ **f** $\dfrac{3\frac{5}{8} - 2\frac{2}{3}}{4\frac{1}{6} + 3\frac{1}{2}}$

3 Use the power, cube and cube root keys on your calculator to calculate these. Round your answers to 1 decimal place if necessary.

a 3^7 **b** $\sqrt[3]{29}$ **c** $9.8^2 \times 2.5$

d $14^2 \div 15^2$ **e** $\sqrt{1 + 1.8^3}$

4 Solve each of these problems using the fraction key of your calculator.

a A wheel turns seventeen $\frac{1}{4}$ turns clockwise, then three $\frac{2}{5}$ turns anticlockwise, then seven $\frac{2}{3}$ turns clockwise. How far has it turned altogether?

b A cat eats $\frac{2}{5}$ of a tin of cat food every day. How long would $5\frac{1}{2}$ tins last?

c Salami costs £8.58 per kilogram. How much does $3\frac{3}{8}$ kg cost?

d A giant company Christmas cake is cut into six equal parts. Two-fifths of one part is divided between eight office staff.

    i What fraction of the cake does each person receive?

    The remaining cake is divided between twenty-five factory workers.

    ii What fraction of the cake does each factory worker receive?

    iii What is the difference between the amount received by an office worker and a factory worker?

Practice

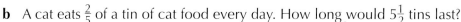

8F Multiplying and dividing decimals

Do not use a calculator.

1 Calculate these.

a 0.3×0.9	**b** 1.7×0.8	**c** 0.7×0.04
d 0.08×0.08	**e** 5.2×9.3	**f** 3.8×6.6
g 4.31×2.7	**h** 0.49×0.78	

2 Calculate these.

a $6 \div 0.3$	**b** $12 \div 0.04$	**c** $0.8 \div 0.02$
d $1.8 \div 0.06$	**e** $2.16 \div 0.4$	**f** $0.24 \div 1.6$
g $29.9 \div 23$	**h** $1.12 \div 0.35$	

3 Bricks are 23.7 cm long. How far would 17 bricks reach laid end to end?

 4 The table shows the postage needed for some letters.

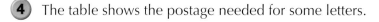

		First-class	Second-class
Format	**Weight**	**Price**	**Price**
Letter	0–100 g	£0.34	£0.24
Large letter	0–100 g	£0.48	£0.36
	101–250 g	£0.70	£0.53

Answer these questions by working in decimals.

a What is the cost of sending 14 letters each weighing 90 g by first-class mail?

b How many second-class 45 g letters could be posted for £4.80?

c Which is cheaper to post: 18 first-class letters each weighing 130 g or 26 second-class letters each weighing 85 g?

Geometry
and Measures **3**

5

Practice

9A Congruent shapes

1 Use your ruler to check which of these triangles are congruent. Write your answer like this, for example, A = D.

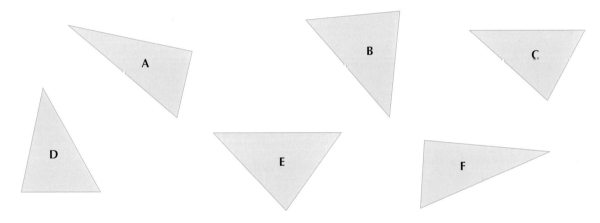

2 Which shapes are congruent? Write your answer like this, for example, A = D = F.

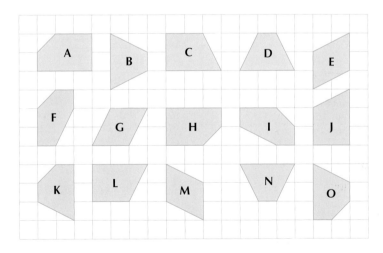

3 a Use triangular dotted paper. Draw 15 shapes by joining some of the dots as shown in the diagram. Label your shapes from A to N.

b Write down the shapes that are congruent.

6

1　**a**　Describe the single transformation that maps the following.
　　i　A onto F
　　ii　D onto A
　　iii　C onto A
　　iv　E onto F
　　v　D onto B
　b　Describe the combination of two transformations that maps the following.
　　i　A onto D
　　ii　B onto C
　　iii　E onto G

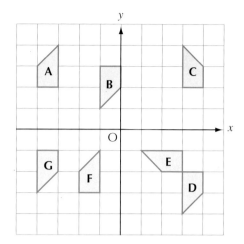

2　**a**　Copy this diagram onto square paper.

　b　Reflect triangle A in the dotted mirror line and label the image B.

　c　Reflect triangle B in the x-axis and label the image C.

　d　What single transformation maps shape A onto shape C?

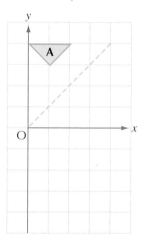

3　**a**　Reflect shape A in the y-axis and label the image B.
　b　What combination of two transformations maps shape A onto shape B?

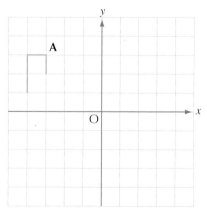

4　**a**　Which single transformation maps a shape onto itself?
　b　Describe a combination of two transformations that map a shape onto itself.

Practice

9C Enlargements

1 Trace each shape with its centre of enlargement O. Enlarge the shape by the given scale factor.

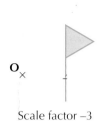

Scale factor –3

Scale factor –2

2 Copy each diagram and enlarge the shape about the point C using the given scale factor.

a

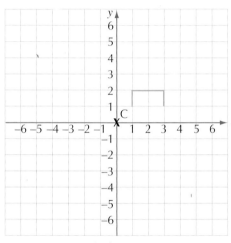

Scale factor –2

b

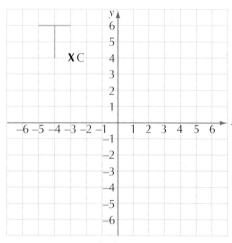

Scale factor –4

3 Shape A'B'C'D' is the enlargement of shape ABCD using a scale factor of –2. Find the points A, B, C and D using rays and draw the shape ABCD.

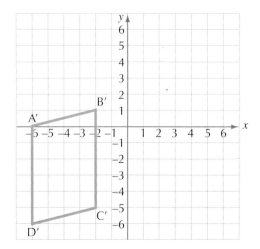

7

1 Write down the number of planes of symmetry that each shape has.

a

3 cm
3 cm

b

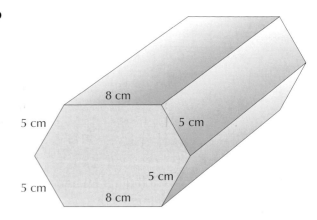

8 cm
5 cm
5 cm
5 cm
5 cm
8 cm

2 Sketch a solid with exactly one plane of symmetry. Use dotted lines to show the plane of symmetry.

3 Copy each shape using an isometric grid. Draw a plane of symmetry using dotted lines or shading. Make a copy of the shape for every plane of symmetry.

a

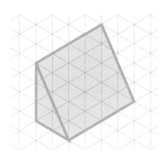

b

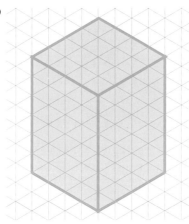

c

5

1 Express each of these ratios in its simplest form.

a 90 cm : 20 cm   **b** 32 mm : 72 mm   **c** 150 cm : 2 m
d 1.8 cm : 40 mm   **e** 0.65 km : 800 m

2

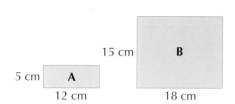 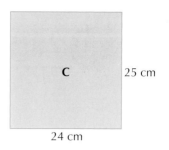

15 cm   **B**

5 cm   **A**

12 cm   18 cm

C   25 cm

24 cm

a Find the ratio of the base of rectangle A to rectangle B.
b Find the ratio of the area of rectangle A to rectangle B.
c What fraction is area A of area B?
d Find the ratio of the area of rectangle B to rectangle C.
e Which is greater: area A as a fraction of area B or area B as a fraction of area C?

FM **3** This diagram shows the design of a new flag.

a Calculate the ratio of the white area to the blue area.
Use ratios to answer these questions.

b 60 m² of white cloth is used to make some flags.
How much blue cloth is needed?

c The total area of some flags is 200 m².
  i How much white cloth do they contain?
  ii How much blue cloth do they contain?

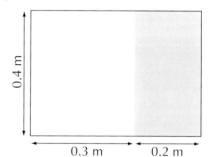

0.4 m

0.3 m   0.2 m

FM **4** This diagram shows the plan of a garden.

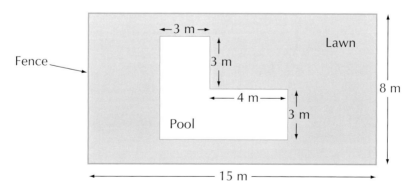

←3 m→
Fence
3 m
Lawn
←4 m→
3 m
8 m
Pool
15 m

a Find the ratio of the perimeter of the fence to the perimeter of the pool.
b Calculate the area of the pool.
c Calculate the area of the lawn.
d Find the ratio of the lawn area to the pool area.

CHAPTER 10 Algebra 4

10A Solving equations

1 Solve these equations.

   a $3a + 7 = 13$ **b** $2n - 3 = 11$ **c** $5v - 20 = 30$
   d $6x + 5 = 29$ **e** $10c - 17 = 73$ **f** $19 = 2y + 13$

2 Solve these equations.

   a $8 + 3t = 23$ **b** $6 + 5m = 16$ **c** $9 + 4d = 29$
   d $13 + 8s = 101$ **e** $43 + 2z = 51$ **f** $63 = 7 + 7i$

3 Solve these equations. The answers are decimals or fractions.

   a $4q + 1 = 15$ **b** $5h - 3 = 3$ **c** $10y + 21 = 69$
   d $3f - 2 = 5$ **e** $8e + 5 = 19$ **f** $5 = 5w - 9$

4 Solve these equations. Expand the brackets first.

   a $2(n + 1) = 12$ **b** $5(p - 2) = 25$ **c** $4(g + 3) = 44$
   d $3(2r + 1) = 15$ **e** $2(3b - 7) = 4$ **f** $70 = 5(4i + 2)$

5 Solve the equations in Question 4 again. This time, remove the brackets by dividing both sides by a number.

6 Solve these equations. Use any method you like.

   a $4r - 20 = 8$ **b** $9 = 5x + 2$ **c** $4(y + 4) = 24$
   d $3 + 5a = 48$ **e** $3(5h - 8) = 36$ **f** $7 + 4u = 25$

7 Helmut's homework and solutions are shown below. Explain what is wrong with each solution. Find the correct answer, where necessary.

Questions	Solutions
a $9x - 2 = 25$	**a** $9x - 2 = 25 + 2 = 3$
b $3 + 4y = 19$	**b** $3 + 4y = 19$ $4y = 19 + 3 = 22$ $y = 5.5$
c $2(2x - 3) = 14$	**c** $2(2x - 3) = 14$ $4x - 3 = 14 \Rightarrow 4x = 17$ $x = 4.25$

10B Equations involving negative numbers

Solve these equations.

1 **a** $3x + 11 = 2$ **b** $2y + 15 = 3$ **c** $4a + 6 = 2$ **d** $15 = 5C + 40$

2 **a** $12 + 3x = 6$ **b** $6 + 5b = 31$ **c** $22 + 6m = 4$ **d** $36 = 100 + 16q$

3 **a** $4d - 19 = -3$ **b** $2x + 7 = -3$ **c** $3z - 14 = -32$ **d** $-22 = 5t - 12$

4 **a** $-4i = 28$ **b** $-3H = 6$ **c** $-8k = -16$ **d** $30 = -6x$

5 **a** $6 - 4x = 18$ **b** $1 - 7u = 15$ **c** $3 - 3d = -12$ **d** $16 = 8 - 2m$

6 Expand the brackets first.

    a $4(s + 2) = 4$ **b** $3(m - 5) = -6$
    c $2(3n + 9) = 6$ **d** $-10 = 5(2y - 12)$

7 **a** $-5x = 10$ **b** $8 - 5x = -12$ **c** $12 + 5f = -3$
    d $2(x + 3) = -10$ **e** $-1 = 3w - 7$ **f** $4(2d - 1) = -36$

10C Equations with unknowns on both sides

Solve these equations.

1 **a** $5y = 9 + 2y$ **b** $10x = 20 + 6x$
    c $8u = 3u + 25$ **d** $7p = 3p + 32$

2 **a** $20 - 3x = x$ **b** $12 - 2c = 2c$
    c $4d = 30 - d$ **d** $5p = 14 - 2p$

3 **a** $6g + 5 = 2g + 13$ **b** $9i + 7 = 5i + 19$
    c $10h - 3 = 3h + 18$ **d** $8t - 15 = 6t + 3$

4 **a** $35 + 2k = 9k$ **b** $3x + 8 = 5x$
    c $2x + 23 = 7x + 3$ **d** $r + 16 = 5r - 10$

5 These equations involve negative numbers.

    a $8y = 6y - 10$ **b** $6k - 6 = 9k$ **c** $40 + 7j = 2j$
    d $5d + 9 = 3d + 3$ **e** $-4r = 3r + 21$ **f** $7n - 3 = 3n - 15$

6 Expand the brackets first.

    a $3(x + 2) = 2x + 8$ **b** $3(2g + 3) = 4g + 17$
    c $5(s - 2) = 3(s + 4)$ **d** $8w - 12 = 3(3w - 1)$

10D Substituting into expressions

1 Write down the value of each expression for each value of x.

a $3 + 4x$ when:
   i $x = 2$ ii $x = 7$ iii $x = -3$

b $9p - 2$ when:
   i $p = -4$ ii $p = 0$ iii $p = 8$

c d^2 when:
   i $d = 7$ ii $d = -6$ iii $d = -1$

d $2s^2$ when:
   i $s = 3$ ii $s = 10$ iii $s = -2$

e $m^2 - 5$ when:
   i $m = 4$ ii $m = 1$ iii $m = -4$

f $5(3n - 2)$ when:
   i $n = 2$ ii $n = 0$ iii $n = -5$

2 If $p = 3$ and $q = 5$, find the value of each of these expressions.

a $2q - p$ b $q + 3p$ c $p^2 + q^2$
d $4(3p - q)$ e $2q - (4p - q)$

3 If $r = -2$ and $s = 3$, find the value of each of these expressions.

a $2r + s$ b $r - 2s$ c $3(2r + 3s)$
d $r - (2s + 1)$ e $10s + 2(r - s)$

4 If $m = 2$ and $n = -4$, find the value of each of these expressions.

a $n^2 - m^2$ b $5m^2$ c mn^2
d $5mn - m^2$ e $m^2 - (2m - n)$

5 If $x = 3$, $y = 5$ and $z = -2$ find the value of each of these expressions.

a xyz b $xz + y$ c $2x - 3y - 4z$
d $(x - z)(x + y)$ e $xz + y(2x - z)$

10E Substituting into formulae

1 The ideal weight, W kg, of a man of height h cm is given by this formula:

$W = 0.75h - 67.5$

Calculate the ideal weight of a man of the following height.
a 160 cm b 186 cm c 1.7 m

2 Given that $C = F + 4d$, find the value of C when:

a $F = 12, d = 6$ b $F = 200, d = 65$ c $F = 18, d = -3$

3 The cost of a sheet of glass is given by the formula $C = 3bh$
where C is the cost (£), b the breadth and h the height in metres.

Calculate the cost of these sheets of glass.
a   Breadth 2 m, height 4 m
b   Breadth 3 m, height 1.5 m
c   Breadth 0.3 m, height 0.8 m

4 Given that:
$$W = \frac{2m + 2n + p}{2}$$

calculate the value of W when:

a $m = 4$, $n = 2$, $p = 6$
b $m = -3$, $n = 2$, $p = 5$
c $m = 6$, $n = -1$, $p = -4$

5 The approximate area of a circle of radius r is given by the formula $A = 3r^2$ where A is the area and r is the radius.

Find the area of a circle of the following radius.
a   5 cm   b   2.5 cm   c   0.9 cm

Practice

10F Creating your own expressions and formulae

1 Write an expression for each of these, using the letters suggested.

a   The sum of the numbers d and 3.
b   The number u reduced by 5.
c   The product of the numbers w, 7 and s.
d   One third of the number m.
e   The number of toes on f feet.

2 How many metres are there in x centimetres?

3 Michael is H cm tall now.

a   He was 5 cm shorter a year ago. How tall was he then?
b   In 3 years' time, he will be x cm taller. How tall will he be then?
His taller sister Briony has a height of S cm.
c   How much taller is Briony?
d   What is the average height of Michael and Briony?

4 a   A Chocolate Wheel costs c pence and a Frother costs f pence.
       What is the total cost of the following?
       i   A Chocolate Wheel and a Frother
       ii 5 Frothers
       iii 3 Chocolate Wheels and 2 Frothers
   b   How much change from a £2 coin would you receive for each of the purchases in **a**?
   c   A packet of Fruit Spots costs 5p more than Frothers.
       i   What is the total cost of 3 packets of Fruit Spots?
       ii How many packets of Fruit Spots can be bought for £2?

      iii How much more do 2 Chocolate Wheels cost than a packet of
            Fruit Spots?

      iv There are 10 Fruit Spots in a packet. How much does each Fruit
            Spot cost?

  d 4 Frothers cost 40p more than a packet of Fruit Spots. Find the cost of
     a Frother.
     (Hint: Write an equation and solve it.)

5 Kevin spent T pence on text messages one week. Marit spent 30p more than
Kevin. Akemi spent twice as much as Kevin.

  a How much did they spend altogether?

  b If they spent 270p altogether, how much did Kevin spend?
     (Hint: Write an equation and solve it.)

6 Mr Walker's daughter is x years old. Five years ago, Mr Walker was twice as
old as his daughter and their ages added up to 57.

  a What were their ages five years ago?

  b What was the total of their ages five years ago?

  c Find the age of Mr Walker's daughter.
     (Hint: Write an equation and solve it.)

CHAPTER 11 Statistics 2

Practice

11A Statistical surveys

 Conduct a statistical investigation to test one of the hypotheses below.
Your investigation must include a questionnaire.
*   Write down three or four good questions for your questionnaire.
*   Make a data recording sheet.
*   Collect information from at least 30 people.

1 Girls eat more fruit than boys.

2 Pupils prefer more English lessons or more Maths lessons.

3 Year 7 pupils watch less football than Year 8 pupils.

4 People visit the cinema once a year, on average.

5 People think crime has increased in their area.

Practice

11B Stem-and-leaf diagrams

1 The average speeds of 35 racing cars are shown in this stem-and-leaf diagram.

11	0	9	9	9								
12	1	2	2	3	4	4	5	6	6	6	8	9
13	0	4	4	5	5	6	6	7	8	8	8	8	9
14	0	3	3	6	8	9						

Key:
14|3 means
143 km/h

a What is the mode?
b Find the median.
c Calculate the range.
d The qualifying speed for the next race is 136 km/h.
How many cars qualified?

2 The ages of 23 children are shown in this stem-and-leaf diagram.

11	9	9	9	10	11	11						
12	0	0	1	2	2	3	4	5	7	10	10	10
13	0	3	3	4	5							

Key:
12|4 means
12 years and 4 months

(Note: Months are rounded down to the nearest month.)
a How many children are 12 years old?
b How many children are younger than $12\frac{1}{2}$?
c Calculate the mode, median and range.

3 These are the weights of 27 tomatoes in a box (in grams).

48	29	62	48	96	45	30	59	16
88	63	10	74	64	38	90	25	8
44	92	67	78	33	50	23	60	18

a Make a stem-and-leaf diagram for the data. Remember to show the key.
b State the mode.
c Calculate the range.
d Find the median.
e Tomatoes weighing 60 g or more are called large.
How many large tomatoes are there in this box?

7

1 Draw pie charts to represent the data.

a The numbers of birds spotted on a field trip.

Bird	Crow	Thrush	Starling	Magpie	Other
Frequency	19	12	8	2	19

b The sizes of dresses sold in a shop during one week.

Size	8	10	12	14	16	18
Frequency	3	7	10	12	6	2

2 150 children went on one of four
school summer holidays.
How many children chose these
holidays?

a Camping
b France
c Pony trekking
d Disney World

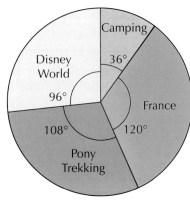

FM **3** The grouped bar chart shows the
average number of text messages people make using their mobile phones
during a month.

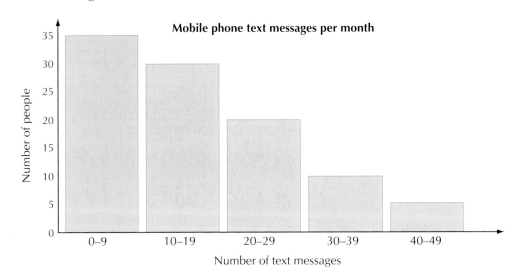

a *Mobile Weekly* magazine claims that 50% of people make more than
25 text messages per month. Could this be true?
b Is it possible that more people make no text messages than those who
make 30 text messages or more?
c Is it true to say that the most frequent number of text messages made is
between 0 and 9, inclusive?
d Write down two more observations based on the diagram.

FM **4** Chin Lin asked 20 people which instrument they prefer to listen to.
She drew this pie chart to show the results.

Favourite instrument

a Roughly half the people chose two of the instruments. Which two?

b Sean says that at least 5 people prefer to listen to the saxophone. Is he correct?

c The violin is more popular than the bagpipes and saxophone combined. Is this true?

d Just over twice the number of people chose the piano compared to another instrument. Which instrument?

e Stringed instruments are twice as popular compared to wind instruments. Is this true? (Hint: Stringed instruments include the piano.)

f Write down two more comparisons based on the pie chart.

g Chin Lin made two mistakes in her diagram. What are they?

Practice

11D Scatter graphs

1 This table shows the class positions of pupils in their Science and Mathematics classes.

Pupil	Jo	Ken	Lim	Tony	Dee	Sam	Pat	Les	Kay	Val	Rod
Science	24	3	13	21	5	15	22	7	12	27	14
Maths	20	5	10	14	9	15	26	1	16	24	11

a Draw a scatter graph for the data.
Use the *x*-axis for Science position, from 0 to 30.
Use the *y*-axis for Mathematics position, from 0 to 30.

b Describe what the graph tells you.

c Which of these best describes the degree of correlation: positive correlation, negative correlation, no correlation?

2 This table shows the heights and History test scores of 20 children.

Height (cm)	130	110	160	120	120	140	160	150	110	130	140	120	170	150	140	160	110	150	140	140
Score	13	19	12	14	7	20	14	10	9	3	17	20	20	15	8	6	13	20	6	12

a Draw a scatter graph for the data.
Use the *x*-axis for height, from 100 cm to 170 cm.
Use the *y*-axis for history test score, from 0 to 20.

b Describe what the graph tells you.

c Describe the degree of correlation.

Practice

11E Analysing data

FM For each question below, do the following.
* Write a hypothesis.
* Decide how best to collect the data (questionnaire, experiment, secondary sources, for example, books, newspapers).
* Make a data recording sheet. Collect the data.
* Organise your data into tables, etc.
* Draw diagrams to illustrate your data.
* Make any necessary calculations, for example, mean, range.
* Write a conclusion.

7

1 How many times can you catch a ball with your weaker hand before dropping it?

2 Do we eat more vegetables now, compared to 10 years ago?

3 How often do people eat from a barbecue?

4 Are the leaves at the bottom of a plant bigger than the leaves in the middle?

5 How long does it take the average person to travel to work?

CHAPTER 12 Number 4

Practice

12A Fractions

5

1 Copy and complete these.

a $\dfrac{9}{4} = \dfrac{\square}{12}$

b $\dfrac{20}{3} = \dfrac{180}{\square}$

c $\dfrac{11}{2} = \dfrac{99}{\square}$

d $\dfrac{32}{7} = \dfrac{\square}{35}$

e $\dfrac{100}{28} = \dfrac{\square}{7}$

f $\dfrac{144}{18} = \dfrac{24}{\square}$

2 **a** How many fifths are in $3\frac{4}{5}$?

  b How many thirds are in $7\frac{1}{3}$?

  c How many twelfths are in $4\frac{7}{12}$?

  d How many sevenths are in 32?

3 Write each of these as a mixed number in its simplest form.

  a $\frac{13}{5}$ **b** $\frac{32}{3}$ **c** $\frac{28}{9}$

  d $\frac{40}{6}$ **e** $\frac{60}{8}$ **f** $\frac{84}{18}$

  g $\frac{133}{21}$ **h** Thirteen thirds **i** Eighteen eighths

4 Write each of these as a fraction.

  a The fraction of a litre given by each of the following.
    i 320 cl **ii** 645 cl **iii** 272 cl **iv** 9375 ml

  b The fraction of a foot (ft), where 12 inches (in) = 1 foot, given by each of the following.
    i 20 in **ii** 51 in **iii** 100 in **iv** 114 in

5 Convert each of these fractions to an improper (top-heavy) fraction.

  a $1\frac{4}{5}$ **b** $2\frac{1}{3}$ **c** $4\frac{3}{10}$ **d** $33\frac{1}{3}$ **e** $5\frac{2}{7}$

Practice

12B Adding and subtracting fractions

If necessary, convert your answers to mixed numbers and cancel down.

1 Use an eighths fraction line to calculate these.

  a $\frac{7}{8} + \frac{3}{4}$ **b** $1\frac{5}{8} + 2\frac{7}{8}$ **c** $1\frac{1}{4} - \frac{3}{8}$ **d** $3\frac{1}{2} - 1\frac{5}{8}$ **e** $2\frac{1}{4} + 1\frac{3}{8} - 2\frac{5}{8}$

2 Calculate these.

  a $\frac{2}{7} + \frac{3}{7}$ **b** $\frac{11}{12} - \frac{7}{12}$ **c** $\frac{4}{5} + \frac{3}{5} + \frac{4}{5}$ **d** $\frac{5}{8} + \frac{7}{8} - \frac{3}{8}$

3 Convert the fractions to equivalent fractions with a common denominator. Then calculate the answer.

  a $\frac{1}{5} + \frac{1}{2}$ **b** $\frac{1}{10} + \frac{2}{5}$ **c** $\frac{5}{6} + \frac{4}{9}$

  d $\frac{3}{8} + \frac{5}{6} + \frac{1}{4}$ **e** $\frac{3}{4} - \frac{1}{3}$ **f** $\frac{11}{12} - \frac{3}{4}$

  g $\frac{7}{9} - \frac{1}{3}$ **h** $\frac{9}{10} - \frac{1}{2} - \frac{1}{5}$

4 Convert the mixed numbers to improper (top-heavy) fractions. Then calculate the answer.

  a $2\frac{3}{5} + 1\frac{1}{2}$ **b** $1\frac{5}{6} + 3\frac{1}{4}$ **c** $4\frac{2}{3} - 1\frac{3}{4}$ **d** $3\frac{3}{8} - 1\frac{1}{6}$

5 Of Jan's emails, $\frac{3}{5}$ is junk mail and $\frac{1}{10}$ is from friends. The rest is work related.

    a What fraction is work related?

    b Jan received 120 emails during the week. How many were *not* junk mail?

6 A poster is printed using red, blue and yellow inks. Of the ink used, $\frac{2}{9}$ is red and $\frac{1}{6}$ is blue.

    a What fraction of the ink used is *not* yellow?

    b 36 ml of ink is used to print the poster. Calculate the amount of each ink used.

7 Sharon bought a bag of flour weighing $3\frac{3}{4}$ kg. She used $\frac{7}{10}$ kg for a cake and $1\frac{3}{5}$ kg for some bread.

    a How much flour did she use altogether?

    b How much flour did she have left over?

Practice

12C Order of operations

Do not use a calculator. Show all of your working.

1 Write down the operation that you would do first in each calculation. Then calculate the answer.

    a $12 - 3 \times 2$ **b** $2 \times (9 - 5)$ **c** $10 \times 2 \div 5 + 3$

    d $30 - 20 + 10$ **e** $12 + 8 - 3^2$ **f** $4 \times (2 + 5)^2$

2 Calculate these. Show each step of your calculation.

    a $3^2 + 5 \times 2$ **b** $10 - (1 + 2)^2$ **c** $3 \times 12 \div 3^2$

    d $32 \div (3^2 - 1)$ **e** $\dfrac{60 + 12}{2 \times 3}$ **f** $\dfrac{60}{(6 + 3^2)}$

    g $1.5 + 3 \times (2.4 - 0.8) - 2.1$

3 Copy each calculation. Insert brackets to make the answer true.

    a $11 - 7 - 1 + 4 = 1$ **b** $1 + 4 + 3^2 = 50$

    c $24 \div 2 \times 3 = 4$ **d** $6 + 9 \div 12 \div 4 = 5$

    e $12 - 3^2 - 7 \times 4 = 4$

4 Calculate these. Work out the inside brackets first.

    a $120 - [70 - (90 - 55)]$ **b** $[(2 + 7)^2 - 12] \div 3$

    c $(2 + 6) \times [36 \div (9 - 5)]$ **d** $12^2 - [5^2 - 2 \times (32 - 27)]$

Practice

Practice

12D Multiplying decimals

Do not use a calculator.

1 Calculate these.

    a 200×0.9 **b** 0.8×400 **c** 500×0.09
    d 2000×0.7 **e** 0.006×600 **f** 70×0.04
    g $90\,000 \times 0.004$ **h** 3000×0.002 **i** 0.0004×400

2 Calculate these.

    a $300 \times 0.4 \times 0.8$ **b** $0.006 \times 7000 \times 0.2$
    c $0.04 \times 0.07 \times 300$ **d** $30\,000 \times 0.002 \times 0.04$

3 Calculate these.

    a 0.6×0.7 **b** 0.9×0.9 **c** 0.05×0.7 **d** 0.8×0.03
    e 0.4^2 **f** 0.04×0.09 **g** 0.003×0.6 **h** 0.008×0.002

4   **a** 0.24×7.5 **b** 0.46×0.32 **c** 12.1×0.18 **d** 23.4×11.7

5 A seed weighs 0.04 g. How much does a box of 600 seeds weigh if the box alone weighs 17 g?

6 Sound travels about 0.3 km in 1 second. How far does sound travel in the following times? Work in kilometres.

    a 200 seconds   **b** 10 minutes   **c** 0.02 seconds

7 Ham costs £8.31 per kg. How much does 0.18 kg cost? Round your answer to the nearest 1p.

Practice

12E Dividing decimals

Do not use a calculator.

1 Calculate these.

    a $0.6 \div 0.03$ **b** $0.08 \div 0.5$ **c** $0.36 \div 0.04$ **d** $0.9 \div 0.02$
    e $0.04 \div 0.001$ **f** $0.15 \div 0.002$ **g** $0.07 \div 0.05$ **h** $0.8 \div 0.002$

2 Calculate these.

    a $9 \div 0.3$ **b** $40 \div 0.8$ **c** $60 \div 0.015$ **d** $48 \div 0.12$
    e $500 \div 0.025$ **f** $900 \div 0.003$ **g** $5000 \div 0.02$ **h** $120 \div 0.24$

3 Calculate these.

  a 2.8 ÷ 20 **b** 32 ÷ 800 **c** 1.2 ÷ 300
  d 0.16 ÷ 400 **e** 0.54 ÷ 90 **f** 0.008 ÷ 400

4 Calculate these.

  a 35.2 ÷ 1.1 **b** 11.7 ÷ 2.6 **c** 3.12 ÷ 4.8 **d** 13.23 ÷ 2.7

5 0.008 g of platinum costs £0.16. Calculate the cost of 1 g. Work in pounds.

6 A website gives 3.6p to charity on every sale it makes. In one month, the website gives £86.40 to charity. How many sales did it make?

7 **a** 3000 bacteria have a mass of 0.000 003 g. What is the mass of one bacterium? (Bacteria is the plural of bacterium.)
  b 10 000 000 000 000 atoms have the same mass as one bacterium. What is the mass of an atom?

CHAPTER **13** Algebra **5**

Practice

13A Expand and simplify

For Questions 1–2, simplify these expressions.

1 **a** $7s + 2s + 4t$ **b** $9i - 4i + 2j$ **c** $4a + 3b + 2b$
  d $5d + 4y + 3d$ **e** $6r + 3h - 4r$ **f** $5b + 3d - 7d$
  g $5u + 2a + 3u + 4a$ **h** $4d + 5y - 2y + 3d$
  i $10k + 4p - 7k - 2p$ **j** $6t - 3g + 2t - 4g$

2 Expand these expressions.

  a $4(x + 8)$ **b** $7(3d - 5)$ **c** $2(4f + 2e)$
  d $d(3 - u)$ **e** $m(2r + c)$ **f** $j(3h - 2g)$

For Questions 3–4, expand and simplify these expressions.

3 **a** $2x + 3(x + 5)$ **b** $4d + 2(5d + 6)$ **c** $9i - 4(i + 2)$
  d $7u - 3(2u - 3)$ **e** $6k + 2e + 3(2k + e)$ **f** $4a + 7b - 2(5a - 2b)$

4 **a** $3(4i + 2) + 4(i + 2)$ **b** $3(2s + 3) + 2(s - 4)$
   c $2(3i - 5) + 3(4i - 1)$ **d** $4(2s + 3) - 3(s + 2)$
   e $5(2f + 2) - 3(3f - 3)$ **f** $4(3 + 2g) - 3(2 - 4g)$

5 Solve these equations.

   a $2(d + 1) + 3(d + 2) = 23$ **b** $3(2s + 5) + 4(s - 3) = 43$
   c $3(3w + 2) - 2(2w - 1) = 28$ **d** $2(2k + 3) - 3(k + 1) = 1$

13B Solving equations by trial and improvement

1 Solve the equation $x^2 + x = 80$, giving your answer correct to 1 decimal place. Copy and complete this table.

x	$x^2 + x$	Comment
8	$8^2 + 8 = 72$	too small
9		
8.5		
8.4		
8.45		
8.46		

2 Solve these equations, giving your answer correct to 1 decimal place. Make a table for each equation.

   a $x^2 + x = 40$ **b** $x^2 - x = 66$ **c** $x^3 + x = 7$ **d** $x^3 - x = 20$

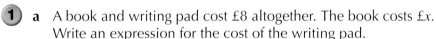

13C Constructing equations

1 **a** A book and writing pad cost £8 altogether. The book costs £x.
      Write an expression for the cost of the writing pad.
   b Maria is twice the age of Janine. Janine is j years old.
      Write an expression for the age of Maria.
   c The difference between the heights of two trees is 4 m. The taller tree is T metres high. Write an expression for the height of the shorter tree.

2 Solve each of these problems by creating an equation and then solving it.

   a Tyres cost £x each. 4 tyres cost £112.
      Find the cost of one tyre.
   b Jamie is y years old and Paul is 9 years old. Their ages total 16 years.
      How old is Jamie?
   c The difference between two numbers is 5. The smaller number is n and the larger number is 12.
      What is the smaller number?

d A plum weighs p grams. A lemon weighs 12 g more than the plum.
  i Write an expression for the total weight of the lemon and plum.
  ii If the total weight is 44 g, find the weight of the plum.

e A soap opera lasts 3 times as long as a cartoon. The cartoon lasts m minutes. The soap opera and the cartoon last 32 minutes altogether.
  i Write an expression for the total length of the programmes.
  ii The soap opera and the cartoon last 32 minutes altogether. How long does the cartoon last?

f The sum of three consecutive numbers is 78. The smallest number is n. Find the value of n.

g I am x years old. Three times my age 4 years ago is 27. How old am I?

h A bakery sold 15 boxes of cakes. Large boxes contain 4 cakes and small boxes contain 3 cakes. 51 cakes were sold altogether.
  How many large boxes were sold?
  (Hint: Let the number of large boxes be x.)

Practice

13D Problems with graphs

1 Find the gradient of each of these lines.

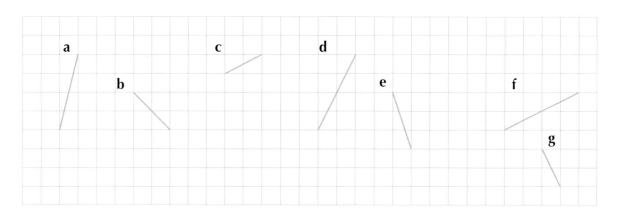

2 Find the gradient of the straight line that joins each pair of coordinates. Plot the points using squared paper.

  a (1, 4) and (3, 10)   **b** (2, 10) and (4, 2)   **c** (0, 3) and (6, 6)
  d (−3, 2) and (0, 8)   **e** (−1, 6) and (1, −6)

3 Find the gradient and y-axis intercept for each of these equations.

  a $y = 3x + 5$ **b** $y = -2x + 3$ **c** $y = 0.5x - 3$ **d** $y = 3x$

4 Write the equation of each line in the form $y = mx + c$, given this information.

  a $m = 5$, $c = 2$ **b** Gradient is 6, y-intercept is −3
  c Gradient is −3, y-intercept is 9   **d** Gradient is 1.2, y-intercept is 0

5 Find the gradient, *y*-intercept and equation for each graph below.

a

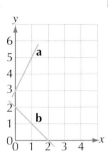

b

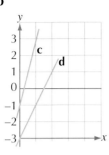

c

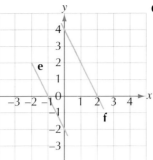

d

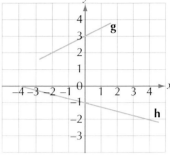

Practice

13E Real-life graphs

1 Sketch a graph to illustrate each of these situations. Estimate any necessary measurements.

    a The height of a person from birth to age 30 years.
    b The temperature in summer from midnight to midnight the next day.
    c The height of water in a WC cistern from before it is flushed to afterwards.

 2 This diagram shows the cost of car hire for two companies.

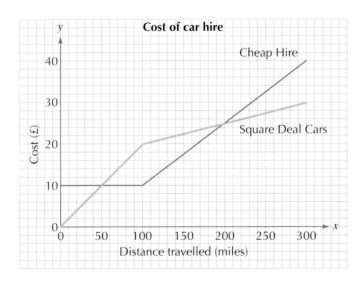

    a Estimate the cost of hiring a car from each company and travelling the following distances.
      i 60 miles   **ii** 100 miles   **iii** 240 miles
    b Describe the way each company charges for car hire.
    c When is Cheap Hire cheaper than Square Deal Cars?

3 Match each description to its graph.

  a The temperature of the desert over a 24-hour period.
  b The temperature of a kitchen over a 24-hour period.
  c The temperature of pond during a month of winter.
  d The temperature of a cup of tea as it cools down.

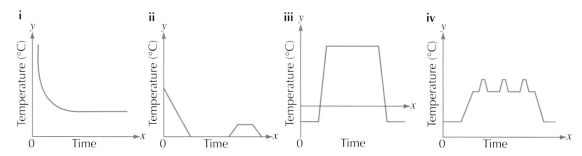

FM **4** George tried several diets to lose weight. This table shows his weight loss over a one-year period. He weighed 105 kg at the beginning of the year.

Diet	Weight loss/gain
Hay diet	Lost 10 kg in 2 months
Calorie counter	Stayed same weight for 3 months
Vegetarian	Gained 3 kg in 2 months
Usual diet	Gained 12 kg in 1 month
Atkins diet	Lost 28 kg in 4 months

Draw a graph showing how George's weight changed over time.
Start the vertical weight axis at 80 kg and use a scale of 1 cm to 2 kg.

Practice

13F Change of subject

1 Rewrite each of these formulae as indicated.

  a $T = D + 4$ — Express D in terms of T.
  b $m = \frac{d}{5}$ — Express d in terms of m.
  c $P = 6T$ — Express T in terms of P.
  d $R = 7 - s$ — Express s in terms of R.

2 Rewrite each of these formulae as indicated.

  a $A = 9p$ — Make p the subject of the formula.
  b $y = 2x + 5$ — Make x the subject of the formula.
  c $A + 5 = 2C$ — Make C the subject of the formula.
  d $y = 8 + c$ — Make c the subject of the formula.
  e $T = 4(m - 5)$ — Make m the subject of the formula.

3 The perimeter of a shape is given by the formula $P = 4a + 5$.

    a Find the value of P when $a = 4$ cm.
    b Make a the subject of the formula.
    c Calculate the value of a when $P = 37$ cm.

4 The speed of a car is u mph. It accelerates to a speed of v mph. Its final speed v mph is given by the formula $v = u + 16$.

    a Calculate the final speed of the car if it accelerates from 20 mph.
    b Make u the subject of the formula.

5 $A = \dfrac{a + 4}{2}$

    a Find A when $a = 5$.
    b Make a the subject of the formula.
    c $A = 18$. Use your formula to find a.

 CHAPTER **14** Solving Problems

Practice

14A Number and measures

1 Two pens and a pencil have a total length of 49 cm. Four pens and a pencil have a total length of 77 cm.
What is the total length of seven pens and a pencil?

2 Copy and complete this puzzle so that each large triangle adds up to 50.

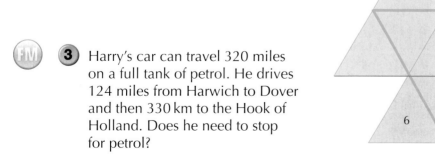

FM 3 Harry's car can travel 320 miles on a full tank of petrol. He drives 124 miles from Harwich to Dover and then 330 km to the Hook of Holland. Does he need to stop for petrol?

4 **a** Find three consecutive even numbers that add up to 114.
    b Find two consecutive even numbers with a product of 1088.

5 A bag of potatoes in the UK weighs 15 lb. A bag of potatoes in France weighs 7 kg. Which is heavier?

6 Two 12-hour electric clocks start at 12 noon. Clock A loses 5 minutes every hour. Clock B keeps perfect time.

   a What will be the time showing on clock A when clock B next reaches 12 o'clock?
   b How long will it take before both clocks show 12 o'clock again?

14B Using algebra, graphs and diagrams to solve problems

1 A pond is 240 cm deep. The depth of the pond decreases 2 cm every day.

   a How long before the pond is half its original depth?
   b Write a formula to show the depth D cm of the pond in n days.

2 I think of a number, treble it and then subtract 24. The answer is 27. What was the number I first thought of?

3 Five times the number I am thinking of is 16 more than treble the number. What is the number I am thinking of?

4 A fruit bowl weighs 500 g and holds oranges that weigh 200 g each.

   a Write a formula to show the total weight W of n oranges.
   b Draw a graph to show the total weight of the fruit bowl and oranges. Use the x-axis for Number of Pieces of Fruit, numbered from 0 to 7. Use the y-axis for Weight, numbered from 0 to 2000 g.
   Another fruit bowl weighs 800 g and holds apples that weigh 100 g each.
   c Draw another graph *using the same axes* to show the total weight of a bowl of apples.
   d Use your graphs to find when a bowl of oranges weighs the same as a bowl of apples.

5 **a** Fold an A4 sheet of paper into three equal parts. Fold the folded sheet in half. Open the sheet and count the rectangles.
   b Repeat part **a** but, this time, fold the folded sheet in half twice.
   c Write down a formula for the number N of rectangles made by folding the folded sheet in half n times.
   d How many rectangles would be made by folding the A4 sheet 5 times?
   e How many times would you have to fold the sheet to make 200 rectangles or more?

14C Logic and proof

1 Copy and complete these number problems.

a
```
  ☐ 5 ☐
+ 1 ☐ 7
───────
  4 4 6
```

b
```
    3 ☐
×   ☐ 4
───────
  8 8 8
```

c $7\,☐^2 = ☐\,9\,☐\,9$

d $(10 + ☐)(☐ + 8) = 182$

2 a Give an example to show that the product of three odd numbers is odd.
   b Prove that the product of three odd numbers is odd.
   (Hint: Remember the product of two odd numbers.)

3 a Give an example to show that the product of two odd numbers with two even numbers is even.
   b Show that the product of four numbers – two odd and two even – is even.

4 a List the factors of 21.
   b Show that all of the factors of an odd number are odd.

5 It takes 8 people 12 hours to dig a hole. How long would it take the following numbers of people?

   a 4 people   b 2 people   c 6 people

6 6 glasses hold 120 cl altogether. 8 mugs hold 144 cl altogether. Which holds the most, a mug or a glass?

7 Which of these numbers must be odd? Explain your answer using algebra.

   a $6n$ b $4n - 1$ c $2n + 5$ d $n + 3$

   (Hint: n is any positive integer.)

8 Use algebra to prove that subtracting an even number from 20 leaves an even number.

9 Use algebra to prove that multiplying an odd number by 3 gives another odd number.

10 Use algebra to prove that squaring an even number gives an even number.

14D Proportion

1 Two of the carriages on a train are first class. The other six carriages are second class.

 a What proportion of carriages are first class?
 b What is the ratio of first-class to second-class carriages?
 A train with 12 carriages has the same proportion of first-class carriages.
 c How many first-class carriages does it have?

2 The ratio of nitrogen to oxygen in the air is approximately 4 : 1.
 A cupboard contains 600 litres of air.

 a How much nitrogen does it contain?
 b How much oxygen does it contain?

3 8 light bulbs cost £7.20. How much does a box of 20 light bulbs cost?

4 There are three females for every four males in a social club with 91 members. How many females are there?

5 8 grams of silver are used to make 15 cm of chain.

 a How much silver does 21 cm of chain contain?
 b How long is a chain that contains 25 grams of silver?

6 A photo of area 108 cm^2 is enlarged to have an area of 180 cm^2.
 What is the ratio of the two areas?

7 12 kg of a new metal alloy was made using this formula:

Tin	400 g
Copper	3 kg
Lead	1800 g
Iron	6.8 kg

 Calculate the formula to make 21 kg of the metal alloy.

14E Ratio

1 **a** Express these ratios in the form n : 1.
   i 18 : 3   **ii** 16 : 10   **iii** 9 : 15   **iv** 400 : 125
 b Express the same ratios in the form 1 : n.

2 **a** Divide 40 cm in the ratio 7 : 1.
 b Divide 600 mm in the ratio 11 : 9.

c Divide 5000 people in the ratio 3 : 5.
d Divide £76 in the ratio 2 : 7 : 10.

3 Donna eats three times more hot meals than cold meals. She eats 96 meals in July. How many of them were hot?

4 The ratio of children in Year 8 who own a mobile phone to those who do not is 3 : 2. There are 210 children in Year 8.
How many children do *not* own a mobile phone?

5 The land area to sea area of the surface of the Earth is in the ratio 3 : 7.
The total surface area of the Earth is 500 000 000 km².
What is the area of land?

6 Marcia receives 12 items of junk mail for every 5 letters. Rajid receives 9 pieces of junk mail for every 4 letters.

a Write the ratio of junk mail to letters for each person in the form *n* : 1.
b Who receives the greater proportion of junk mail?

CHAPTER 15

Geometry and Measures **4**

Practice

15A Plans and elevations

1 Draw each of these 3-D shapes on an isometric grid.

a

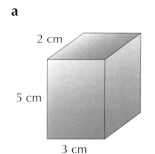

2 cm
5 cm
3 cm

b

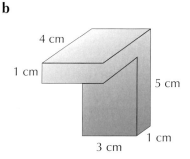

4 cm
1 cm
5 cm
3 cm
1 cm

c

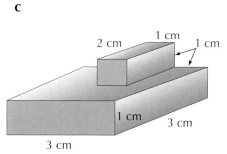

2 cm
1 cm
1 cm
1 cm
3 cm
3 cm

2 For each of these 3-D shapes, draw a **i** plan **ii** front elevation **iii** side elevation.

a **b**

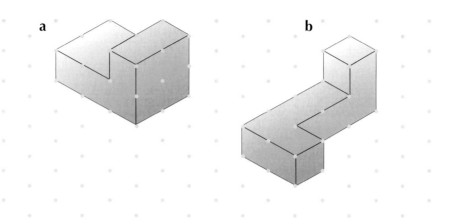

3 The plan, front and side elevations of a 3-D shape are shown below. Draw the solid on an isometric grid.

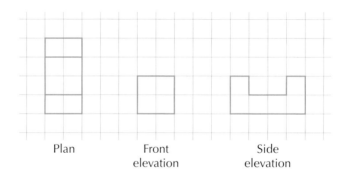

Plan   Front   Side
elevation   elevation

4 The diagram shows two views of the same solid.

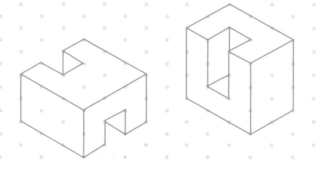

Draw enough plan and elevation diagrams for someone to follow to make the shape out of plastic cubes.

1 These objects have been drawn using the scales shown. Find the true lengths of the objects.

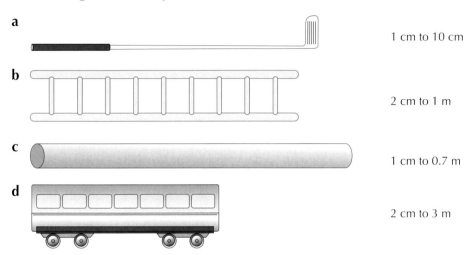

a 1 cm to 10 cm

b 2 cm to 1 m

c 1 cm to 0.7 m

d 2 cm to 3 m

  2

Scale: 1 cm to 120 m

The diagram shows a scale drawing of an aircraft hangar.

a Calculate the real length of the hangar.
b Calculate the real width of the hangar.
c Calculate the area of the hangar.

3 Copy and complete this table.

	Scale	Scaled length	Actual length
a	1 cm to 2 m		24 m
b	1 cm to 5 km	9.2 cm	
c		12 cm	42 miles
d	5 cm to 8 m	30 cm	

CH 15

6

FM 4 This diagram shows a scale drawing of a supermarket.

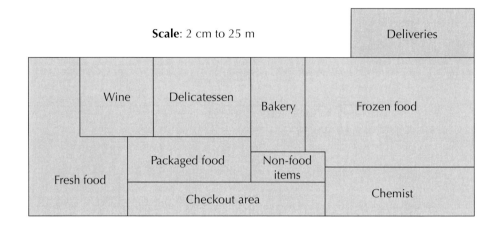

Make a table showing the dimensions and area of each section of the supermarket.

Practice

15C Finding the mid-point of a line segment

1 Copy this grid and plot the points.

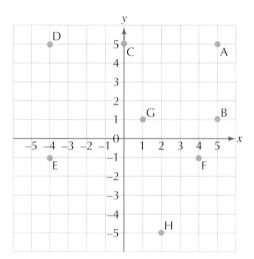

a List the coordinates of the points.
b Join these pairs of points using straight lines.
   i AB   ii AD   iii DE   iv EF
   v AG   vi EH   vii BE
c Find the mid-point of each line in **b**.
d List the coordinates of the mid-points.

2 Copy this grid and plot the points.

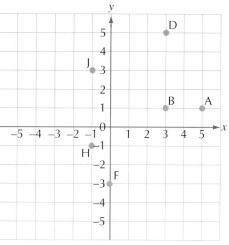

a B is the mid-point of line AC.
Find the coordinates of point C.
b B is the mid-point of line DE.
Find the coordinates of point E.
c F is the mid-point of line EG.
Find the coordinates of point G.
d J is the mid-point of line DK.
Find the coordinates of point K.
e The origin is the mid-point of KL.
Find the coordinates of point L.

3 Find the mid-points of these line
segments. Use any method you know (diagram or calculation).

a A(1, 3) joined to B(5, 7)   **b** C(0, 4) joined to D(5, 8)
c E(−4, 0) joined to F(0, 6)   **d** G(−6, −2) joined to H(2, −4)

Practice

15D Map scales

1 Write each of these map scales as a map ratio, that is, 1 : number.

a 1 cm to 3 km   **b** 1 cm to 100 m   **c** 2 cm to 100 km

 2 This map shows an area of England and Wales. Find the shortest (straight line) distance between these pairs of places. Give your answers to the nearest 5 km.

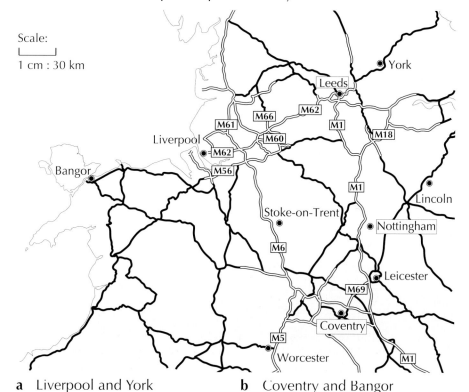

Scale:
1 cm : 30 km

a Liverpool and York   **b** Coventry and Bangor
c Worcester and Leeds   **d** Lincoln and Stoke-on-Trent

3 The distance between Knutsford and Littleborough is 42 km.
How far apart would these towns be on the above map?

4 Barry made a journey using these motorways.
Nottingham to Leicester on the M1
Leicester to Coventry on the M69
Coventry to Stoke-on-Trent on the M6

Use a piece of string on the above map to estimate the total length of his journey, to the nearest 10 km.

5 The ratio of a map is 1 : 200 000.

   a Two places are 15 cm apart on the map. What is their actual distance apart?
   b Two places are 71 km apart. How far apart are they on the map?

15E Loci

1 Sketch loci for these. Describe each locus.

   a A moving garden swing
   b A moving ski lift chair

2 This diagram shows three wooden balls in a game of bowls.
Trace the diagram.

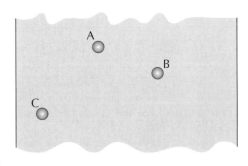

   a A fourth ball, D, moves equidistant from balls A and B. Sketch its locus.
   b A fifth ball collides with the balls. Ball C moves so that it is equidistant from the edge of the bowling green. Sketch its locus.

3

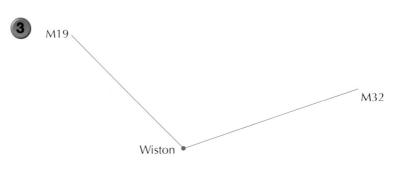

Stan drives his car from Highpoint, heading eastwards, and keeps the same distance from Wiston. When Stan is equidistant from the motorways, he drives away from Wiston, keeping equidistant from the motorways. Trace the diagram and sketch the locus of Stan's car.

4 Mark a point on the circumference of a 2p coin. Move the coin along the edge of a ruler. Draw the locus of the marked point.

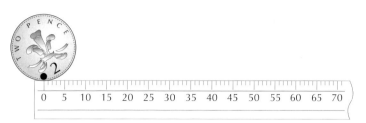

5 The wheel continuously runs along this channel, staying in contact with the edge Y.

a Trace the diagram.
b Draw the locus of the centre of the wheel.

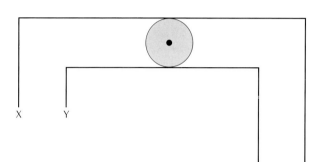

X   Y

15F Bearings

1 Use your protractor to find the bearing of B from A.

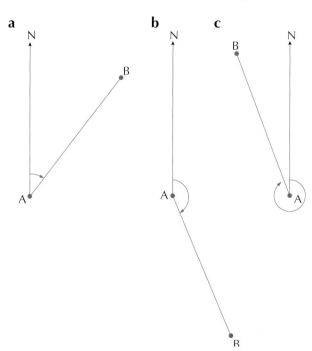

a

N

B

A

b

N

A

R

c

B

N

A

2 Find the bearing of P from Q.

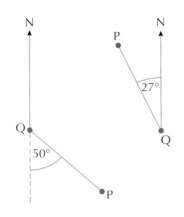

3 Sketch these bearings. Label your diagrams.

a A skier is on a bearing of 72° from the ski lodge.
b Clerkhill is 12 miles from Moffat on a bearing of 115°.

4

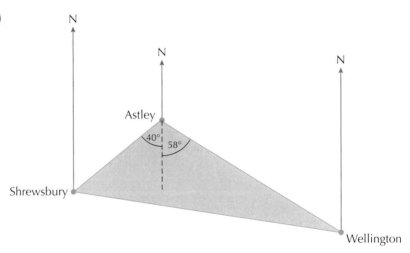

Calculate the bearing of:

a Wellington from Astley   **b** Astley from Wellington
c Astley from Shrewsbury   **d** Shrewsbury from Astley

Use your protractor to find the bearing of:

e Wellington from Shrewsbury   **f** Shrewsbury from Wellington

CHAPTER 16 Statistics 3

1 This table shows the lengths (*L* metres) of 30 python snakes.

Length of snake (*L* metres)	Frequency
$3.5 \leqslant L < 4$	2
$4 \leqslant L < 4.5$	4
$4.5 \leqslant L < 5$	7
$5 \leqslant L < 5.5$	9
$5.5 \leqslant L < 6$	5
$6 \leqslant L < 6.5$	3

a One of the snakes is 4.5 m long. Which class contains this length?
b How many snakes are shorter than 5 metres?
c How many snakes have a length of 5.5 metres or more?
d How many snakes are between 4 and 6 metres long?
e Why is your answer to **d** an approximation?

2 The volumes (*V* cl) of liquid contained in 20 coconuts are shown below.

12.2	11.1	10.5	12.8	12.0	10.1	11.8	12.3	10.7	12.7
10.0	11.6	12.1	10.5	10.8	12.6	10.7	11.4	12.8	11.3

Copy and complete this table.

Volume of liquid (*V* cl)	Tally	Number of coconuts
$10 \leqslant V < 10.5$		
$10.5 \leqslant V < 11$		
$11 \leqslant V < 11.5$		
$11.5 \leqslant V < 12$		
$12 \leqslant V < 12.5$		
$12.5 \leqslant V < 13$		

3 The durations (*t* minutes) of 30 telephone calls are shown below. Times have been rounded up to the nearest minute.

4	12	8	1	19	7	7	28	14	54
9	2	20	16	2	43	5	18	1	5
5	14	9	10	3	30	11	6	17	2

Construct a grouped frequency table for the data.

4 Describe in detail how you would collect data to investigate each of these.

   a The bounce of a ball dropped from different heights.
   b Most people find that alternative medicine doesn't work.
   c It is cheaper to use public transport than to travel by car.

16B Assumed mean and working with statistics

1 **a** Find the mean of 256, 251, 259, 250, 255, 255, 253, 249. Use 250 as assumed mean.
   b Find the mean of 72.7, 72, 72.9, 71.7, 72.8, 72.3, 72.6, 72.2, 72.6, 72.2. Use 72 as assumed mean.
   c Find the mean of 58 284, 58 280, 58 287, 58 278, 58 289. Use 58 280 as assumed mean.

2 Write down two numbers with a range of 4 and a mean of 7.

3 Write down three numbers with a mode of 10 and a mean of 9.

4 The mean of four numbers is 5, the mode 1 and the range is 11. What are the four numbers?

5 The mode of a set of numbers is 5, their range is 10 and their mean is 20. Each of the numbers is increased by 10.

   a How are the averages affected?
   b What is the new range?

 6 The weekly wages of four employees are £320, £290, £420 and £370.

   a Calculate the mean and range for the wages.
   b The employer hires a fifth worker. She wants the mean wage to be £340. What should she pay the fifth worker?
   c All workers received a pay rise of £30 per week. How will this affect the mean and range? (Do not recalculate them.)
   d The next year, the workers receive a 10% pay rise. How do you think the mean and range will be affected?

16C Drawing frequency diagrams

1 Draw a bar chart for each frequency table.

   a Amounts of beer served in 500 ml glasses.

Amount of beer (V ml)	Number of glasses
$485 \leqslant V < 490$	3
$490 \leqslant V < 495$	7
$495 \leqslant V < 500$	13
$500 \leqslant V < 505$	18
$505 \leqslant V < 510$	11

b Shop prices for the same beach ball.

Price of beach ball (£P)	Number of shops
£2.80 $< P \leqslant$ £3	27
£3 $< P \leqslant$ £3.20	51
£3.20 $< P \leqslant$ £3.40	30
£3.40 $< P \leqslant$ £3.60	23
£3.60 $< P \leqslant$ £3.80	13
£3.80 $< P \leqslant$ £4	9

2 This table shows the average height of a sweet corn bush after being sprayed with different amounts of a new additive.

Amount of additive (A ml)	Height of plant (h m)
0	1.40
10	1.40
20	1.45
30	1.55
40	1.70
50	1.70
60	1.60
70	1.50
80	1.40
90	1.30
100	1.30
110	1.30

a Plot a graph. Use these scales.
x-axis (Amount of additive): 1 cm to 10 ml
y-axis (Height of plant): 2 cm to 0.10 m
b Estimate the level of additive that would give plants of height 1.65 m.
c Estimate the height of a plant sprayed with 75 ml of additive.
d Which level of additive would you advise the farmer to use?
Explain your answer.
e At which levels did the additive not improve growth?

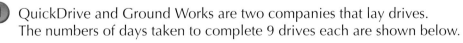

Practice

16D Comparing data

1 QuickDrive and Ground Works are two companies that lay drives.
The numbers of days taken to complete 9 drives each are shown below.

QuickDrive   2, 5, 3, 3, 6, 2, 1, 8, 3

Ground Works   3, 2, 3, 1, 4, 3, 2, 2, 1

a Calculate the mode, mean and range for each company.
b Comment on the differences between the averages.
c Comment on the difference between the ranges.

2 This table shows the weights of fish (grams) that three anglers caught in a competition.

Jerry	Aditya	Marion
230	230	100
100	400	130
380	280	430
720	320	200
450	250	70
		180
		200
		90

a Calculate the mean and range for each of the anglers.
b Who was the most consistent? Explain your answer.
c Who performed the best overall? Explain your answer.
d Which angler would you choose to enter a competition that offered prizes for the heaviest fish caught? Explain your answer.

 3 This table shows the mean and range for the average weekly rainfall (mm) in two holiday resorts.

	Larmidor	Tutu Island
Mean	6.5	5
Range	33	62

Explain the advantages of each island's climate using the mean and range.

Practice

16E Comparing sets of data

1 These pie charts show how two companies first contacted new customers.

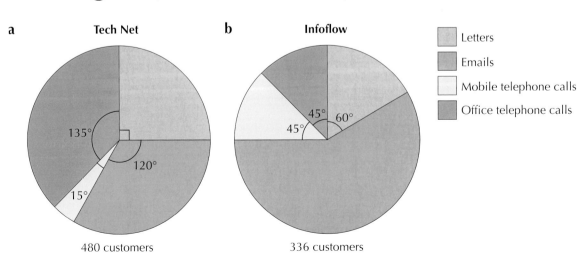

a **Tech Net**
135°
15°
120°
480 customers

b **Infoflow**
45° 60°
45°
336 customers

Letters
Emails
Mobile telephone calls
Office telephone calls

Explain your answers to these questions.

a Which company sent the most emails?
b Which company made the most telephone calls?
c Make two more comparisons between the companies.

2 This bar chart shows the lengths of holidays sold by two shops.

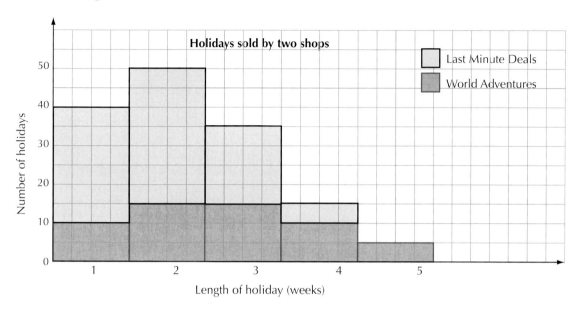

Write a report comparing the sales of the two shops.

3 This graph shows the number of salmon caught at a fishing lodge during 1990 and 1995.

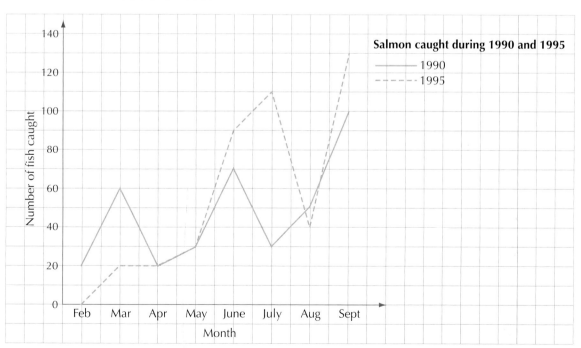

a Make a table of the data.
b Write a report comparing the catches during 1990 and 1995.

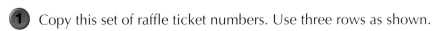

7

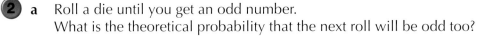

1 Copy this set of raffle ticket numbers. Use three rows as shown.

1	2	3	4	5	6	7	8	9	10
11	12	13	14	15	16	17	18	19	20
21	22	23	24	25	26	27	28	29	30

a What is the theoretical probability of someone choosing a number between 11 and 20 (inclusively)?

b Design and carry out an investigation to test this hypothesis:
People are most likely to choose a number from the middle row.

2 a Roll a die until you get an odd number.
What is the theoretical probability that the next roll will be odd too?

b Design and carry out an experiment to test this hypothesis:
An odd number is more likely to be followed by an even number than by an odd number.

Your Notes

Your Notes